Organizational Behaviour in Health Care

Series editors
Jean-Louis Denis
Ecole Nationale d'Administration
Université de Montréal
Montreal, QC, Canada

Justin Waring
Centre for Health Innovation Leadership and Learning
Nottingham University Business School
Nottingham, UK

Paula Hyde
Manchester Business School
University of Manchester
Manchester, UK

Published in co-operation with the Society for Studies in Organising Healthcare (SHOC), this series has two strands, the first of which consists of specially selected papers taken from the biennial conferences held by SHOC that present a cohesive and focused insight into issues within the field of organisational behaviour in healthcare.

The series also encourages proposals for monographs and edited collections to address the additional and emergent topics in the field of health policy, organization and management. Books within the series aim to advance scholarship on the application of social science theories, methods and concepts to the study of organizing and managing healthcare services and systems.

Providing a new platform for advanced and engaged scholarship, books in the series will advance the academic community by fostering a deep analysis on the challenges for healthcare organizations and management with an explicitly international and comparative focus.

More information about this series at
http://www.springer.com/series/14724

Aoife M. McDermott · Martin Kitchener
Mark Exworthy
Editors

Managing Improvement in Healthcare

Attaining, Sustaining and Spreading Quality

Editors
Aoife M. McDermott
Cardiff Business School
Cardiff University
Cardiff, UK

Mark Exworthy
Health Services Management Centre
University of Birmingham
Birmingham, UK

Martin Kitchener
Cardiff Business School
Cardiff University
Cardiff, UK

Organizational Behaviour in Health Care
ISBN 978-3-319-62234-7   ISBN 978-3-319-62235-4   (eBook)
https://doi.org/10.1007/978-3-319-62235-4

Library of Congress Control Number: 2017948302

Printed on acid-free paper

This Palgrave Macmillan imprint is published by Springer Nature
The registered company is Springer International Publishing AG
The registered company address is: Gewerbestrasse 11, 6330 Cham, Switzerland

Foreword

This book brings together a strong collection of chapters grouped around *Managing Improvements in Healthcare*. It addresses questions about how to attain, embed and sustain improvements in healthcare organisation and delivery. Each chapter reflects the challenges of and opportunities for achieving improvement across various international health systems. This book is presented in three parts; the first covers aims and approaches in quality improvement and examines different perspectives on quality through systematic studies in various international health contexts. The second concerns how to spread and embed quality improvements, including via an examination of various strategies for knowledge mobilisation. The third part concerns the various agents, co-producers and recipients of quality care. Where this work challenges existing perspectives for both academic and practitioner communities, it offers an up-to-date analysis of academic work and practitioner developments.

Various ideas about quality improvement are studied in a way that connects academic and practitioner communities and provide insights that have the capacity to transform theory into policy and practice. These works demonstrate the impact that academic work in the field can have, through analyses and evaluations taken from academic studies around the world. This marks a turn in the book series towards issues of process, in particular, towards what has been termed the implementation gap.

This tenth book in the *Organizational Behaviour in Health Care* series brings together papers from the 10th Organisational Behaviour in Health Care (OBHC) conference held at Cardiff Business School, Cardiff University, Wales, in April 2016. The title of the conference was 'Attaining, sustaining and spreading improvement', and the conference was hosted by Cardiff Health Organisation and Policy Studies group (CHOPS). The conference was a great success with over 120 delegates from 18 countries across Europe, North America and Australia. We would like to thank Dr. Aoife McDermott and Prof. Martin Kitchener, the members of the scientific committee, and all at Cardiff Business School.

The conference series is organised by the Society for Studies in Organising Healthcare (SHOC), which is a learned society and a member of the UK Academy of Social Sciences. The purpose of SHOC is to '[a]dvance the education of the public in the study of the organisation of health care including the promotion of research and the dissemination of the useful results thereof'. SHOC sets up a scientific committee to plan and oversee each OBHC conference, including local academic partners. We are now looking forward to the 11th OBHC conference to be held in Montreal in April 2018, entitled 'Co-ordinating care across boundaries and borders: Systems, networks and collaborations'.

Paula Hyde
OBHC Series Editor

Contents

Contributors

Comfort Adeosun is a healthcare professional with over two decades of experience in patient care and management. In recent years, her career focus has been on healthcare management and health services research in order to contribute to policymaking and innovation management. Comfort obtained her MBA and Ph.D. in Management Studies from the University of Aberdeen. Her particular interests include qualitative research, quality improvement, organisational development and the implementation of new innovation and interventions in organisations.

Rebecca Amati is an external doctoral candidate of the Center for the Advancement of Healthcare Quality and Patient Safety at the Università della Svizzera italiana (USI Lugano). She works for the Quality Department of a private non-profit clinic in Switzerland.

Simon Bailey is a research fellow at Alliance Manchester Business School. His research examines new knowledge practices, and how they emerge and become embedded in healthcare organisations. He is currently working on research exploring GP federations and on a study of the embedding of new working practices across organisational boundaries.

Margaret Banks is a Program Director for the Australian Commission on Safety and Quality in Health Care, Sydney, Australia.

Emma Barnes is based at the School of Social Sciences, Cardiff University. With a background in qualitative methods, Emma started her research career at Bristol University before moving to Cardiff in 2006, working on a range of healthcare-related projects. Emma joined CUREMeDE as a Research Associate in 2010.

Emilie Berard is a Professor at Instituto Tecnológico y de Estudios Superiores de Occidente (ITESO) in Guadalajara, Mexico, and a member of the Health Management Innovation Center, ESCP Europe. She works on healthcare management issues from a management control perspective. Her research focuses on the implementation process of managerial innovations.

Ruth Boaden is Director of the National Institute for Health Research (NIHR) Collaboration for Leadership in Applied Health Research and Care (CLAHRC) Greater Manchester, and Professor of Service Operations Management at Alliance Manchester Business School, University of Manchester.

Jeffrey Braithwaite is Foundation Director, Australian Institute of Health Innovation; Director, Centre for Healthcare Resilience and Implementation Science; and Professor of Health Systems Research, Faculty of Medicine and Health Sciences, Macquarie University, Australia. His research examines the changing nature of health systems, has resulted in over 630 refereed publications, and has attracted more than AUD$102 million in research funding; he has received numerous national and international awards for his teaching and research.

Mark Brandon is the Chief Policy and Regulatory Officer for a large Australian residential aged care provider, and a former vice-chair of the International Society of Quality in Health Care (ISQua) Accreditation Council. He is former CEO of the Australian Government's Aged Care

Standards and Accreditation Agency. As the founding convenor of the ISQua Quality in Social Care for Older Persons interest group, he has provided consultancy services around the world in the areas of aged care, quality and standards development.

Robert Brook was one of the founders of the quality field and has received numerous awards for making quality part of the public policy agenda. As a medical student, he was the person who coded the 9115 critical incidents that were part of the original Sanazaro and Williamson study.

Alison Bullock is based at the School of Social Sciences, Cardiff University. After establishing a career in researching healthcare education at the University of Birmingham, Alison took up her post as Professor and Director of the Cardiff Unit for Research and Evaluation in Medical and Dental Education (CUREMeDE) in 2009. Her research interests include health services management.

Kath Checkland is a GP and Professor of Health Policy and Primary Care at the University of Manchester. She is Associate Director of the DH-funded Policy Research Unit in Commissioning and the Healthcare System (PRUComm). Her research focuses upon the impact of health policy on the NHS, with a particular focus on how organisations interact and respond to change.

Amanda Crompton is Assistant Professor in Public Policy and Management at Nottingham University Business School. Her research examines aspects of the policy process including decision-making practices, stakeholder engagement and implementation. She is currently examining the application of social movement ideas in existing organisations.

Dori A. Cross is a doctoral candidate in the Department of Health Management and Policy at the University of Michigan School of Public Health. Her work focuses on understanding and promoting organisational changes that improve care coordination and care transitions for complex patient populations. This work spans a focus on implementation and use of health information technology, team-based approaches to care and organisational capacity for innovation.

Deborah Debono is currently a Director of Studies of the Health Services Management programme in the Faculty of Health at University of Technology Sydney (UTS). Deborah's research investigates the influence of context, culture, technology and social relationships on clinicians' and health services managers' practice, quality improvement and patient safety.

Jean-Louis Denis is a Professor of Health Policy and Management at the School of Public Health, Université de Montréal, and Senior Scientist, Health Systems and Innovation at the Research Center of the CHUM (CRCHUM). He holds the Canada Research Chair (Tier I) on governance and transformation of healthcare organisations and systems. His current research looks at health system transformation and reforms, medical compensation and professional leadership and clinical governance.

Helen Dickinson Centre for Public Service Research, University of New South Wales, Canberra. Her expertise is in public services, particularly in relation to topics such as governance, leadership, commissioning and priority setting and decision-making. She is co-editor of the Journal of Health, Organization and Management and Australian Journal of Public Administration programmes.

Joy Furnival works as a quality improvement adviser at NHS Improvement. She is also a Health Foundation Generation Q and Improvement Science Ph.D. Fellow and Chartered Engineer. The study detailed in Chapter 4 was conducted as part of her Ph.D. research as a full-time student at Alliance Manchester Business School which was funded by the Health Foundation.

David Greenfield is Professor and Director of the Australian Institute of Health Service Management, University of Tasmania, Australia. His education and research investigate healthcare complex adaptive systems, strategies to improve health services and the organisation of clinical practice.

Annegret F. Hannawa Ph.D. is Associate Professor of Health Communication at the Università della Svizzera italiana (USI) in Switzerland, where she directs a Center for the Advancement of Healthcare Quality and Patient Safety. She also presides over the ISCOME Global Institute for the Advancement of Communication Science in Healthcare.

Gill Harvey is a Professor of Healthcare Management at Alliance Manchester Business School and Professorial Research Fellow at Adelaide Nursing School. She is currently involved in a number of projects implementing and evaluating knowledge mobilisation initiatives in healthcare. A particular interest is in applying broader organisational and management theories of learning and knowledge management to the study of knowledge translation in healthcare contexts.

Brendan Heck is a chartered physiotherapist who currently works as a consultant in the management of health organisations.

Damian Hodgson is Professor of Organizational Analysis and Deputy Director of the Health Services Research Centre at Alliance Manchester Business School, University of Manchester. His research focuses on issues of power, knowledge, identity and control in complex organisational settings. He leads the Organising Healthcare programme in the Collaboration for Leadership in Applied Health Research and Care in Greater Manchester (CLAHRC-GM).

Anne Hogden is a Research Fellow at the Australian Institute of Health Innovation, Macquarie University, Australia. Her research investigates patient-centred care and decision-making using stakeholder consultation.

Hilary Homans has worked with 47 countries on health and development programmes for DfID and the UN. She has also held academic posts in three countries, most recently as Director of the Centre for Sustainable International Development at the University of Aberdeen. Whilst working with the WHO (Geneva) she co-authored the first *Making Pregnancy Safer Strategy*.

John Humphreys is a project manager in the National Institute of Health Research Collaboration for Leadership in Applied Health Research and Care (NIHR CLAHRC) Greater Manchester. He is interested in exploring the limits of managerial approaches in actual practice and understanding the barriers to the implementation of complex organisational change in healthcare organisations.

Amer Kaissi is a Professor of Healthcare Administration at Trinity University, San Antonio, Texas. He has published on quality of care, patient safety and retail medicine. His new book *Intangibles: The Unexpected Traits of High-Performing Healthcare Leaders* will be published by Health Administration Press in August 2017.

Mary A. Keating is Associate Professor in Human Resource Management in Trinity Business School at Trinity College Dublin, Ireland. An occupational psychologist, her research interests concern cross-cultural management and the management of professional service organisations.

Roman Kislov is a Research Fellow at Alliance Manchester Business School, University of Manchester. His research interests include boundary management, communities of practice, organisational learning and knowledge mobilisation in collaborative contexts. He is currently leading a longitudinal study into the development of knowledge mobilisation strategies over time and a multiple case study of research co-production in multiprofessional project teams.

Marlot Kuiper is a Ph.D. researcher at the Utrecht School of Governance (USG), the Netherlands. She obtained a Master's degree in Research in Public Administration and Organizational Science at the Universities of Utrecht, Rotterdam and Tilburg. Currently, she is working on her Ph.D. project 'Connective Routines', financed by the Dutch Organization for Scientific Research (NWO Research Talent Grant) and supervised by Professor Mirko Noordegraaf (USG) and Dr. Lars Tummers (USG).

Jean Ledger is a researcher based at the Department of Applied Health Research, University College London. She works on projects exploring issues such as the adoption of healthcare innovations, policy-driven system change and service improvement. Her academic interests include knowledge mobilisation, qualitative research methodologies and public management.

Anne McBride is a senior lecturer in Employment Studies at the Alliance Manchester Business School, University of Manchester. Her research interests lie in the broad areas of healthcare human resource management and gender relations at work. She has published on the role of different actors in role redesign, workforce development and service improvement and the implications of changing work practices for individuals, groups and organisations.

Lorna McKee is Emeritus Professor of Management and Health Services Research at the University of Aberdeen. Lorna is a sociologist with a long-standing interest in organisations, and she has led projects focused on organisational change and quality improvement within the healthcare sector. Lorna is currently a Visiting Professor at the University of Cologne and an accredited Executive Coach and Leadership Mentor.

Brian Marshall is a Programme Director and OD Faculty member at Ashridge Business School. He has more than twenty years of leadership and line management experience with British Oxygen, Black and Decker and Unipart Group, before changing career direction to become an OD and Change specialist, initially with the British Civil Service and then at Ashridge.

Miguel Martínez-Lucio is Professor of International HRM at the Alliance Manchester Business School, University of Manchester. The main focus of his research is the changing patterns of rights and regulation within employment relations and human resource management. Much of this work has a comparative and international perspective. The central concern of much of his work is the position and role of regulation and institutions in the context of globalisation, increasing managerialism and socio-economic uncertainty.

Virginia Mumford is an applied health economist with previous experience in clinical medicine and international finance. She was awarded a three year NSW Early to Mid-Career Fellowship to evaluate the implementation of a delirium clinical care standard, and is also working on projects to evaluate the introduction of an electronic medication management system in paediatric hospitals, the role of surgery in refractory epilepsy, and the impact of genetic testing in motor neurone disease.

Stephen Parkin is a qualitative researcher based at the Nuffield Department of Primary Care Health Sciences, University of Oxford. He is currently engaged in an ethnographic study of quality improvement in the NHS. His research interests include harm reduction responses to substance use, the application of social theory to real-world issues and applied ethnography.

John G. Richmond is a Ph.D. candidate at Warwick Business School. John's research is funded by the National Institute for Health Research (NIHR) Collaborations for Leadership in Applied Health Research and Care West Midlands (CLAHRC WM). He is a member of the Canadian College of Health Leaders with 10 years of management experience in public and private healthcare in both Canada and the USA. Most recently, John was a Risk Manager for a Health Authority in Eastern Canada.

Olivier Saulpic is a Professor at ESCP Europe, Paris campus. He is Co-Director of the Health Innovation Management Research Centre.

His research is interested in the effects of management tools on organisational change, especially in the healthcare sector.

Kieran Walshe is Professor of Health Policy and Management at Manchester Business School, and head of the Health Management Group at MBS. He is also a non-executive director of the Christie NHS Foundation Trust, a board member of the UK Health Services Research Network and a member of the US Academy Health International Advisory Board.

Justin Waring is Professor of Organisational Sociology and Associate Dean at Nottingham University Business School. His research examines the social organisation and governance of healthcare services, with a particular focus on the changing organisation and management of professional practices, cultures and institutions.

Wendy Warren MBE is a clinician and manager who has formerly worked as a Director of Nursing in both England and Wales. Wendy is Head of Planning and Civil Contingencies with Aneurin Bevan University Health Board, with additional roles in the anti-radicalisation programme PREVENT and the Nursing and Midwifery Council.

Johanna Westbrook is Director of the Centre for Health Systems and Safety Research, Australian Institute of Health Innovation, Macquarie University. Her expertise is in multi-method evaluation in health informatics and patient safety.

Liz Wiggins is Associate Professor of Change and Leadership at Ashridge, now part of Hult Business School. She has over 20 years' experience leading change at Unilever and as a change and communications consultant across the private and public sectors. Liz is also an executive coach and supervisor to clients across a range of sectors, as well as being part of the NHS pool of accredited coaches.

Philippe Zarlowski is a Professor at ESCP Europe, Paris campus. He is Scientific Director of the Deloitte Chair on Public Service and

Managerial Performance, in partnership with ENA. He conducts and coordinates research on the role of performance management systems and tools in the implementation of public policies and the transformation of organisations, notably in the fields of healthcare and local government.

Series Note

Organizational Behaviour in Health Care Series

Series Editors: Professor Jean-Louis Denis, Université de Montréal, Canada; Professor Paula Hyde, Manchester University, UK; Professor Justin Waring, Nottingham University, UK.

A series of biennial volumes, published in cooperation with the Society for Studies in Organizing Healthcare (SHOC). Each volume is comprised of specially selected papers taken from biennial conferences held by SHOC and presents a cohesive and focused insight into issues within the field of organisational behaviour in healthcare.

The Society's goals are:

'Advance the education of the public in the study of the organization of health care, including the promotion of research and the dissemination of the useful results thereof'.

Titles include:

Annabelle Mark and Sue Dopson (Eds.)
ORGANISATIONAL BEHAVIOUR IN HEALTH CARE
The research agenda

Lynn Ashburner (Ed.)
ORGANISATIONAL BEHAVIOUR AND ORGANISATION STUDIES IN HEALTH CARE
Reflections on the future
Sue Dopson and Annabelle Mark (Eds.)
LEADING HEALTH CARE ORGANIZATIONS

Ann L. Casebeer, Alexandra Harrison and Annabelle Mark (Eds.)
INNOVATIONS IN HEALTHCARE
A reality check

Lorna McKee, Ewan Ferlie and Paula Hyde (Eds.)
ORGANIZING AND REORGANIZING
Power and change in health care organizations

Jeffrey Braithwaite, Paula Hyde and Catherine Pope (Eds.)
CULTURE AND CLIMATE IN HEALTH CARE ORGANIZATIONS

Helen Dickinson and Russell Mannion (Eds.)
THE REFORM OF HEALTHCARE: SHAPING, ADAPTING AND RESISTING POLICY DEVELOPMENTS

Mary A. Keating, Aoife M. McDermott and Kathleen Montgomery
PATIENT CENTRED HEALTH CARE
Achieving coordination, communication and innovation

Susanne Boch Waldorff, Anne Reff Pedersen, Louise Fitzgerald and Ewan Ferlie
MANAGING CHANGE
From health policy to practice

List of Figures

List of Tables

Introduction

This edited volume emanates from the work of members of the Society for Studies in Organising Healthcare (known as SHOC), a UK-based Learned Society with international membership. The core purpose of the Society is to advance the study of the organisation of healthcare, and to promote and disseminate the resulting research findings. As part of its work, the Society organises a biannual conference, which rotates between the UK and other countries. This volume presents selected papers from the 10th International Organisational Behaviour in Healthcare Conference, hosted by Cardiff Business School in 2016.

The conference theme was focused on attaining, sustaining and spreading improvement, as is evident in the title of this volume. This focus reflects an international recognition that achieving and embedding improvement in health systems is difficult—but increasingly necessary due to increased demands, limited resources and enhanced evidence regarding what works and why. Full paper and symposia submissions were received from 18 countries, and reviewed by the Scientific Organising Committee—Professors Catherine Pope, Jean-Louis Denis, Mark Exworthy and Paula Hyde—and 49 peer reviewers. The Society particularly encouraged submissions from Ph.D. and early career colleagues, as well as more established scholars. The participation of early career researchers was kindly supported by the provision of bursaries by the Health Foundation.

The high volume and quality of the submissions received from scholars across disciplines and career stages reflect the vibrant state of quality and improvement oriented research in healthcare. This book serves as a record of the high-quality submissions to the conference. It provides a summary of key current issues within the field, and it affords an opportunity to reflect on the conceptualisation and pursuit of quality. Indeed, debate abounds regarding the most important aims of quality improvement. Quality is multifaceted, as evident in the Institute of Medicine's six dimensions of quality; care should be *efficient, effective, equitable, timely, person-centred* and *safe*. Further, there are myriad approaches to achieving quality. As a result, Part I of the book examines the aims of, and approaches to, quality improvement. Within this section, contributors draw attention to approaches to delivering improvement, the influence of national and organisational context, and governance and organisational strategies to support quality.

Whilst important, starting improvement quality initiatives is insufficient. They must be sustained within organisations, with effective interventions spread across the system. Reflecting this, Part II focuses on embedding and spreading quality. Initially, attention is afforded to the importance of 'unlearning'—ceasing ineffective interventions, and creating space for new initiatives. Thereafter, the role played by routines, in anchoring quality in day-to-day activities is, noted. Subsequent attention is afforded to dialogue with and the participation of staff, to ensure the sustainability of quality initiatives. However, systemic quality alone does not necessitate the embedding of effective initiatives—they must also spread. As a result, attention is given to the design of policy initiatives, processes and accountabilities to support knowledge mobilisation and the upscaling of good practice.

Finally, Part III focuses on the key actors involved in quality improvement: change agents, co-producers and recipients. It gives attention to the work inherent in delivering reform, including the strategies, behaviours and roles adopted by managers and staff in support of quality improvement. Crucially, it also gives attention to patient involvement in quality initiatives, and the extent to which the quality interventions support the production of outcomes valued by service users.

In addition to the core, themes considered—the aims of, and approaches to, quality; how to embed and spread it; and the roles of staff and patients in delivering it—the interdisciplinary contributions evidence the value of a range of methods. These range from survey, to interview, to observational approaches. The predominance of qualitative and exploratory studies means that the chapters have scope to prompt theoretical debate and to influence policy and practice. Next, we provide a summary of the chapters in each themed section of the book.

Part I—Quality Improvement: Aims, Approaches and Context

To begin, Amati et al. examine valued aspects of quality care that serve as aims for quality improvement. They replicate a seminal survey of physicians, which asked them to identify the characteristics of good and poor quality care. In their survey of healthcare managers, Amati et al. find that the importance afforded to dimensions of quality has changed over time, with new aspects also emerging relating to both the process (e.g. guidelines adherence, patient-centeredness) and outcomes of care. Amati et al.'s chapter serves to highlight dimensions of quality that improvement initiatives should target. More broadly, it also emphasises shifts affecting healthcare delivery (e.g. evidence-based practice), and the role of patients within this.

Turning from the target to the process of quality improvement, Wiggins and Marshall consider how to select a change or improvement approach from the myriad available. They draw attention to different assumptions underpinning a range of approaches (e.g. Lean, Appreciative Inquiry), including some which are irreconcilable. They identify five potential responses amongst change agents and leaders. Wiggins and Marshall's chapter is informed by their experiences in providing a leadership development programme for healthcare. They used Appreciative Inquiry to test out their ideas with participants, and to note the benefits of becoming comfortable with using different approaches at different times. Their key contribution is an acknowledgement of the challenges that change leaders face given the array of change and improvement methods available. Supporting leaders to develop competence and confidence in selecting and utilising approach approaches to change is paramount.

Turning to the implementation of quality improvement (QI) processes, Berard et al. illustrate how interplay between actors, the context in which they work and the quality interventions introduced can affect the outcomes achieved. They synthesise existing healthcare literature on QI and context, and supplement this with a key concept from the organisational and management literature. Specifically, they use the concept of 'affordances', which refers to the possibilities for action offered by an object, to help supplement contextual explanations for why different outcomes result from the same intervention. Using an example of a budgeting tool, they illustrate how interventions can constrain and enable interpretations of possibilities. On this basis, they suggest that QI design should be given greater attention and incorporated into research reporting.

The remaining three chapters in Part I consider approaches to improvement. Furnival et al. consider regulatory approaches in four countries. They note the emergence of hybrid models, using deterrence and compliance methods concurrent with softer improvement support. They identify the complexities faced by regulators as they balance punishment with persuasion. Specifically, they find that the roles of the regulator, the resources available to them and the relationships with their regulates are complex.

Complexity can arise not just from the roles pursued, but also from context. Cross provides a review of the contextual factors influencing the success of primary care teams. She reviews the environment, task and technological factors likely to affect their success. The review is premised on recognition that evidence on the effects of team-based care is inconsistent, necessitating an understanding of the context and mechanisms supporting realisation of the intended benefits. Crucially this chapter emphasises that interventions are not introduced in a vacuum. Rather, they need to be deployed in supportive contexts, with appropriate resources to enable realisation. Macro (e.g. regulatory, financial), meso (e.g. governance, working practices) and micro (e.g. individual responses) factors all have scope to influence the impact of interventions. Thus, Cross notes the importance of understanding not just whether interventions work, but when.

The first part of this book concludes with Heck and Keating's consideration of the adoption and implementation of Lean as an approach to improvement in Irish hospitals. Based on document analysis and interviews with Lean experts, they identify a piecemeal approach to implementation. They note the need for systemic and systematic adoption of improvement initiatives, to enhance the likelihood of successful and sustained reform.

Thus, Part I introduces the aims of quality improvement, as well as the approaches that managers' can adopt—and how to choose between them. It acknowledges the importance of context, and the potential for systemic and organisational level interventions in support of quality.

Part II—Embedding and Spreading Quality

Part II explores the ways in which national and local contextual factors shape the nature of the implementation of quality improvement interventions. It does so in six chapters which show the tensions between individual and structural, local and national approaches to current research in this field. The implication of these chapters is the need to enhance and refine the process of embedding and spreading quality, including via research translation.

To begin the section, Richmond calls for a focus on 'unlearning' as well as 'learning'. His literature review offers insights into the evidence base of this previously neglected area of study. He adopts Scott's (cognitive, cultural and political) pillars to illustrate this unlearning process.

Thereafter, Kuiper examines the ways in which specific initiatives (such as checklists), used with the intention of embedding improvements in routines, are not always translated in practice by clinicians. She finds that professional norms and values have a significant impact upon observed variation in their use. As such, checklists should not be seen as coordination devices but as the site for connections between multi-professional routines.

It is not just professions which shape—and are shaped by—QI interventions. McBride and Martínez-Lucio investigate the role of trade unions in service improvement. They observe a lack of involvement of trade unions in three national schemes in England. Given the emphasis on participatory approaches in QI, this contradictory approach is significant as it marginalises the collective dialogue from staff and can so undermine the sustainability of the QI initiative itself.

Greenfield and colleagues consider an alternative way of creating staff awareness and ownership of a policy initiative. They adopt an approach which combines national and local perspectives in their study of icons, as symbols of a policy. Icons, they argue, can be used to promote staff knowledge of, and engagement with, national policy at the local level. Using documentary material to examine this, they consider whether the apparently isomorphic processes are coercive or voluntary.

Thus, the early chapters in this section evidence that embedding and spreading quality initiatives are challenging. Building on this theme, Barnes and colleagues examine the ways in which knowledge translation and mobilisation have largely been shaped as an individual routine and responsibility in Wales. As a consequence, they are not embedded in organisational systems and processes. Recognising the scale of the challenge to become embedded, they identify the barriers which need to be overcome and the enabling factors which might accelerate and sustain this endeavour.

By contrast, Dickinson and Ledger also examine issues of translation but in terms of the organisational mechanisms which accelerate this process in Australia. Their focus is on governance and structure as well as the required cultural change.

The chapters in Part II reveal a breadth of research which challenges established patterns of working, of diffusing innovation and accelerating change. The breadth comes not only from the national context (here, Australia, the Netherlands and the UK) but also from theoretical underpinnings. (That said, qualitative methods tend to dominate such scholarship.) Central to this research is the ways in which boundaries are crossed—knowledge, national/local or professional/disciplinary.

Part III—Agents, Co-producers and Recipients of Quality Care

The third and concluding part of this edited collection directs attention towards people and their actions (agency) within quality improvement initiatives. Drawing from a variety of conceptual frames to label participants as change agents, institutional entrepreneurs, co-producers, champions and recipients, this set of papers shares a common concern for the work inherent in delivering reform, including the strategies, behaviours and roles adopted by managers and staff in support of quality improvement. Crucially, and quite innovatively, it also affords attention to patient involvement in quality initiatives, and the extent to which the quality interventions support the production of outcomes valued by service users.

Crompton and Waring make an important contribution to the emerging steam of research into collaborative and participatory improvement methodologies that are designed to create receptive contexts for change. Previous studies in this vein have tended to present unquestioning accounts of the espoused (stated) 'potential' for social movement strategies to engender 'bottom-up' healthcare improvement. In sharp contrast, this piece surfaces a potentially darker side of social movement activity. Drawing from rich UK case study evidence, Crompton and Waring show how healthcare leaders sought to *build a 'movement for improvement'* by using framing (justification) strategies based on claims about empowering frontline clinicians. Their findings suggest that managers' use of social movement ideas seemed little concerned with fostering bottom-up improvement work, and more a means for reducing resistance to a relatively prescribed top-down improvement framework. Such an unmasking of contemporary developments in healthcare organisation is at once rare, refreshing, troubling and stimulating.

Next up, Checkland et al. apply three concepts from institutional theory—logics (belief systems), institutional entrepreneurs, and work—to explore the implementation of projects designed to improve access to English primary care services. Findings from their case analysis illustrate the conflicting nature of extant logics and provide some interesting examples of the micro-level 'creating' and 'disrupting' work of institutional entrepreneurs involved in the change programme. The authors report, however, that it is far from clear whether this institutional work will, in the longer term, accumulate to deliver the intended wider institutional change. Such findings serve to remind us that whilst new policy directions cannot be initiated without disruptive institutional work, disrupting alone doesn't necessarily lead to desired outcomes.

Kislov and colleagues report an early attempt to explore the temporal dynamics and microprocesses involved in the evolution of *facilitation*, a service improvement approach based on the mobilisation of evidence-based knowledge into clinical practice. Drawing on a longitudinal case study of Chronic Kidney Disease services in primary care organisations, they describe the following three parallel and overlapping microprocesses underpinning the gradual distortion of facilitation over time: (1) prioritisation of (measurable) outcomes over the (interactive) process; (2) reduction of team engagement and (3) erosion of the facilitator role. These findings show how the uncritical and uncontrolled adaptation of facilitation may undermine its promise to positively affect organisational learning processes, and how it may also mask the unsustainable nature of the improvement outcomes captured by conventional performance measurement. Whilst this unmasking of political elements shares an outcome with Crompton and Waring's chapter, facilitation seems to offer a no more guaranteed means of delivering sustained institutional change than did the disruptive institutional work reported in Checkland and colleagues' work.

In a real breath of fresh air, Adeosun and colleagues combine a focus on service users (a stakeholder group that is too often ignored within healthcare improvement research) with a distinctive research setting, Africa (a context rarely reported in healthcare research). Specifically, they use a comparative case study methodology to explore factors that influence the implementation effectiveness of a clinical practice guideline for antenatal care in four healthcare organisations in Nigeria. Their findings illustrate how service users are not passive in the implementation process, but rather are active change agents who influence and help to co-shape implementation effectiveness.

Finally, whilst Hogden and colleagues share the previous chapter's interest in service users, they concentrate on a very different improvement approach (accreditation) and context (Australian residential care facilities). More specifically, Hogden and colleagues' case study research concentrates on the ways that accreditation processes inform how residents manage their healthcare and lifestyle. The findings show that residents' expectation that accreditation ensured standards of quality and safety meant that few residents made use of accreditation assessment information. Moreover, it is widely accepted that regulators and policymakers have found it challenging to translate into standards, aspects of care and service that are a priority to residents, such as the sense of being at home, or of being cared for. As with Adeosun and colleagues' findings from a very different context, Hogden and colleagues demonstrate that there are opportunities for greater engagement with service users in approaches to improving the quality of the care that they receive.

Attaining, sustaining and spreading quality improvement

Taken together, these chapter summaries evidence the wide range of issues being researched by members of the Organisation Behaviour in Healthcare Community. The contributions in the volume demonstrate the wide variety of challenges facing those tasked with conceptualising, planning and delivering quality care. Within this, of particular note is the attention afforded to including service users and understanding factors enhancing their, as well as professional, perceptions of quality. The chapters also provide constructive theoretical and practical insights, in support of researchers and practitioners in the field. Whilst starting initiatives has historically received substantive attention, it is theoretically and practically significant that embedding and spreading initiatives are receiving enhanced research attention. Supporting appropriate 'unlearning' whilst avoiding initiative decay for valued interventions are important in ensuring sustained benefits from quality improvement initiatives. Whilst spreading good practice is desirable for patients, providers and service sustainability, attaining, sustaining and spreading quality improvement in healthcare are likely to remain enduring challenges for practitioners and researchers alike.

Part I

Quality Improvement: Aims, Approaches and Context

1

Evolving Dimensions of Quality Care: Comparing Physician and Managerial Perspectives

Rebecca Amati, Robert H. Brook, Amer A. Kaissi and Annegret F. Hannawa

Introduction

Improving healthcare is a goal across the world. In order to reach this goal, it is necessary to develop criteria, indicators and instruments to assess quality. Nearly fifty years ago, Sanazaro and Williamson noticed

R. Amati (✉)
Università della Svizzera italiana (USI), Lugano, Switzerland
e-mail: Rebecca.amati@usa.ch

R.H. Brook
Pardee RAND Graduate School, Santa Monica, CA, USA

A.A. Kaissi
Trinity University, San Antonio, TX, USA

A.F. Hannawa
Università della Svizzera italiana (USI), Lugano, Switzerland

© The Author(s) 2018
A.M. McDermott et al. (eds.), *Managing Improvement in Healthcare*,
Organizational Behaviour in Health Care,
https://doi.org/10.1007/978-3-319-62235-4_1

that not much work had focused on the development of objective criteria of performance (Donabedian 1966). For this reason, they conducted a study to create a classification—based on episodes of care provided by physicians—of what constitutes effective and ineffective performance (Sanazaro and Williamson 1970).

Since that time, a vast amount of literature has been published to understand better what quality care is and to find the most appropriate criteria and tools for its measurement and improvement (Arah et al. 2006; Brook et al. 1996; Campbell et al. 2000; Donabedian 1988, 1990; Institute of Medicine 2001; World Health Organization 2006). Major trends that have originated in the management field—such as Total Quality Management, Quality Assurance, Continuous Quality Improvement, Lean or Six Sigma—have also been applied to healthcare. In addition, publications such as those from the Institute of Medicine 1999, 2001), and associations such as the Joint Commission International, the American Society for Quality, the National Association for Healthcare Quality, the International Society for Quality in Health Care and the Agency for Healthcare Research and Quality have emphasized quality problems and their improvement.

Given this 'quality revolution' (Maguard 2006), we replicated Sanazaro and Williamson's (1970) design about fifty years later, using a sample of healthcare managers, to compare our results to their suggested classification, identifying differences and similarities between physician and managerial perspectives and discussing the evolution of quality dimensions over time.

Methods

This study is part of a larger project (Amati et al. in preparation) to develop an empirically informed taxonomy of quality of care, grounded in Donabedian's structure, process and outcome framework (Donabedian 1996, 1998). We refer to that paper (Amati et al. in preparation) for a more detailed description of the methods used.

We replicated a revised version of the critical incidents technique adopted by Sanazaro and Williamson (1970), who collected 9115 episodes of patient care—describing effective and ineffective performance—from 2342 physicians. Our sample comprised 236 top managers in executive positions, middle managers and directors, who had completed the Masters of Science in Healthcare Administration programme at Trinity University (San Antonio, Texas) from 2004 to 2013. Sanazaro and Williamson's (1970) classification system first divided quality statements into process (i.e. what physicians do to patients) and outcome (i.e. effects of physicians' performance on patients). In addition, they identified specific subcategories of both process and outcome, such as 'arriving at diagnosis' or 'improvement of physical abnormalities'.

Each episode of care from our study was analyzed using this classification, in order to ensure a comparison of the data. Moreover, we used an inductive exploratory approach to examine those parts of the texts that did not belong to any of Sanazaro and Williamson's subcategories, leading to the identification of new dimensions of quality care (Amati et al. in preparation). Finally, after the percentages for each subcategory were calculated, we modified three tables published in Sanazaro and Williamson's (1970) work to compare our results to theirs. The comparison was made by looking at ranks and means and did not use formal statistical analysis.

Results

Sample Characteristics

A total of 135 episodes of care were collected from 74 managers (response rate = 33%). Fifty-three percent of the respondents were female and the average age was 35 years old, with a mean of eight years of experience in healthcare management. Professional titles ranged from 'Executive/Vice President' (24%) and 'Director/Manager' (32%) to 'Assistant/Associate Administrator' (16%) and others, such as 'Consultant' and 'Analyst'. Concerning organizational settings, 56% of the respondents worked in private not-for-profit hospitals, 19% in public hospitals, 17% in private for-profit hospitals, whilst the rest worked

in other types of healthcare organizations (e.g. health insurance compa-
nies or outpatient clinics).

Process Subcategories

Sanazaro and Williamson's Subcategories

Table 1.1 reports the top fifteen process subcategories of effective and
ineffective performance most frequently reported in this investigation,
compared with those from the original work (Sanazaro and Williamson
1970). Overall, Sanazaro and Williamson's (1970) process subcategories
were replicated by our data. However, the ranking and percentages were
quite different from the original study. Since Sanazaro and Williamson's
(1970) investigation used physicians to describe quality of care, their
derived taxonomy was very detailed about certain elements of the delivery of care (e.g. use of instruments, X-ray, EKG, caesarean section, etc.),
which were not as prominent in our study.

Concerning effective performance, seven subcategories appeared in
the top fifteen list of both studies (i.e. *Surgical treatment, Use of facilities, Professional manner, Patient education, Arriving at diagnosis, Drug
treatment* and *Laboratory*). However, some differences could be found:
four subcategories (i.e. *Arriving at diagnosis, Drug treatment, Patient education* and *Laboratory*) were ranked higher by physicians in Sanazaro
and Williamson's work. Furthermore, three additional subcategories
(i.e. *Use of health team, Follow-up* and *Physician availability*) were part
of Sanazaro and Williamson's (1970) overall classification, but did not
belong to their top 15 list, whereas in the eyes of our managers they
assumed more importance. In particular, *Use of health team* was the
most reported subcategory of effective performance in our study.

Concerning ineffective performance, five out of the fifteen subcategories most frequently reported by physicians in Sanazaro and
Williamson's (1970) investigation also belonged to the top fifteen of
our study (*Professional manner, Patient education, Surgical treatment, Use
of facilities* and *Drug treatment*). However, whilst *Professional manner*
was reported more frequently by our managers, *Drug treatment* was

Table 1.1 The top fifteen process subcategories of effective and ineffective performance most frequently reported by the healthcare managers in our study in comparison to Sanazaro and Williamson's study (1970), expressed in percent

| Ranking | Sanazaro and Williamson (1970) | | | | | | | | Our study (2017) | | |
| | Effective | | | | Ineffective | | | | Effective | | Ineffective | |
	Internal med (N = 8521)	Surgery (N = 4100)	Paediatrics (N = 3479)	OBGYN (N = 2166)	Internal med (N = 4059)	Surgery (N = 2272)	Paediatrics (N = 1777)	OBGYN (N = 1221)		Total (N = 333)		Total (N = 229)
1 Arriving at diagnosis	11.7%	11%	11.3%	10.5%	12.4%	8.9%	12.8%	8.7%	Use of health team	13.8%	Staff-patient-family comm.*	15.3%
2 Drugs, biologicals, etc.	9	4	8.6	6.1	11.6	3.2	10.5	7.5	Staff-patient-family comm.*	12.4	Timeliness*	10.5
3 Patient education	6.6	7.4	8.2	9.9	5.6	6.7	7.8	–	Timeliness*	9.3	Inter-staff comm.*	9.2
4 Laboratory	5.8	–	6	–	4.3	–	5.4	–	Patient-centredness*	6.3	Use of health team	8.3
5 Use of facilities	5.3	3.2	6.4	3.8	3.7	–	–	–	Inter-staff comm.*	5.4	Professional manner	7.9
6 General evaluation	5.3	4.3	–	–	4.5	3.3	–	–	Surgical treatment	5.4	Adherence to guidelines*	7.9
7 Surgical treatment	5.3	22.3	4.5	12.6	–	26.1	–	13.5	Use of facilities	4.5	Physician availability	4.8
8 X-ray	5	4.6	7.5	5.8	–	–	4.5	4.8	Professional manner	4.5	Professional responsibility	4.8
9 Physical examination	5	4.6	7.5	5.8	–	–	4.5	4.8	Patient education	4.5	Patient education	4.4
10 Consultation	4.5	3.4	6.2	3.9	4.6	4.8	4.5	3.5	Arriving at diagnosis	3.6	Surgical treatment	3.9
11 Professional manner	–	3.2	–	–	6.4	4.9	6.3	5.7	Adherence to guidelines*	2.7	Consistency/continuity*	3.1

(continued)

Table 1.1 (continued)

| | Sanazaro and Williamson (1970) | | | | | | | | Our study (2017) | | | |
| | Effective | | | | Ineffective | | | | Effective | | Ineffective | |
Ranking	Internal med (N = 8521)	Surgery (N = 4100)	Paediatrics (N = 3479)	OBGYN (N = 2166)	Internal med (N = 4059)	Surgery (N = 2272)	Paediatrics (N = 1777)	OBGYN (N = 1221)		Total (N = 333)		Total (N = 229)
12	History	–	–	4.3	–	–	–	–	–	Follow-up	2.7	Use of facilities	1.7
13	Caesarean section/delivery	–	–	–	7.4	–	–	–	9.5	Drugs, biologicals, etc.	2.4	Drugs, biologicals, etc.	1.7
14	Diagnostic procedures	–	–	–	4.8	–	–	–	–	Physician availability	2.4	Procedure	1.7
15	Psychologic support	–	–	–	4	–	–	–	–	Laboratory	2.4	Follow-up	1.3

*New subcategories under *Our study (2017)*

reported much less frequently than in Sanazaro and Williamson's (1970) work. Five subcategories—which had been identified by Sanazaro and Williamson but that did not belong in their top fifteen list—assumed more salience in our study (i.e. *Use of health team, Physician availability, Professional responsibility, Procedure* and *Follow-up*).

The subcategory *Physician availability* in our investigation included the availability of other healthcare professionals. Overall, in the episodes of care that we collected, five of Sanazaro and Williamson's (1970) top fifteen subcategories appeared as contributors of both effective and ineffective performance (i.e. *Surgical treatment, Use of facilities, Professional manner, Patient education* and *Drug treatment*). *Arriving at diagnosis* and *Laboratory* were amongst the top fifteen only under effective performance, whilst *Professional responsibility* and *Procedure* appeared only under ineffective performance.

New Subcategories

Six new subcategories were identified from our episodes of care (Amati et al. in preparation). Four of them ranked amongst the top fifteen of both effective and ineffective performance: *Staff-patient-family communication, Timeliness, Inter-staff communication,* and *Adherence to guidelines/protocols. Patient-centredness* was a new subcategory under effective performance and *Consistency/Continuity of care* was a new subcategory under ineffective performance.

The subcategory *Inter-staff communication* included more specific communication aspects that were not covered in the subcategory *Use of health team*—which only referred to 'coordinating services of other physicians, nurses, auxiliary workers; promoting, facilitating communication among professionals' (Sanazaro and Williamson 1970, p. 301)—such as handoffs, communicating wrong information, conflict management, alert, documentation, debriefings and 'speaking up'. The subcategory *Staff-patient-family communication* included aspects of *Patient-centredness*—defined by the Institute of Medicine as 'providing care that is respectful of and responsive to individual patient preferences, needs, and values and ensuring that patient values guide all clinical decisions' (2001, p. 40); *Patient education*—i.e. 'instructing, educating; explaining; preparing patients. Primary purpose is increased patient knowledge and understanding of condition and regimen' (Sanazaro and Williamson 1970, p. 302); *Professional manner*—i.e. 'establishing or maintaining rapport; physician behavior/attitudes in dealing with patient' (Sanazaro and Williamson 1970, p. 301); and *Psychologic support*—i.e. 'Reassuring; alleviating concern; expressing interest in patient, family. Goal is improved emotional state' (Sanazaro and Williamson 1970, p. 302).

The above subcategory referred not only to a unidirectional type of communication from the healthcare staff to the patient and the family, but it also emphasized a mutual type of relationship, stressing the importance of the patient 'speaking up', of the quality and timeliness of the information and the manner in which it is exchanged. Furthermore, whilst in Sanazaro and Williamson's (1970) work the subcategory *Patient education* specifically referred to treatment, in our study communication was also about navigating the patient and their family through the healthcare system and the process of care.

If we group all these subcategories under two broad dimensions named *Inter-staff communication* and *Staff-patient-family communication*, the former one would cover 19.2% of all subcategories related to effective performance and 17.5% of all subcategories related to ineffective performance. The latter one would account for 28.8% of all subcategories related to effective performance and 32% of all subcategories related to ineffective performance. Therefore, overall, in this study communication aspects would account for 48% of all subcategories related to effective performance and 49.5% of all subcategories related to ineffective performance.

Outcome Subcategories

Beneficial Outcomes

Table 1.2 reports the top thirteen most frequent beneficial outcomes of our study, compared to Sanazaro and Williamson's (1970). Out of their study's top thirteen beneficial outcomes, six were confirmed in the top thirteen of our results (i.e. *Attitude towards M.D., care: Positive*; *Physical*

Table 1.2 The top thirteen most frequent beneficial outcomes reported in our study, compared to Sanazaro and Williamson (1970), expressed in percent

| Ranking | Beneficial outcomes | | | | Our study (2017) | Total |
| | Sanazaro and Williamson (1970) | | | | | |
	Internal medicine (N = 5554)	Surgery (N = 2804)	Paediatrics (N = 1931)	OBGYN (N = 1248)		(N = 144)
1	Individual function: increased 10.9%	9%	5.6%	4.8%	Attitude towards M.D., care: Positive	14.6%
2	Physical abnormalities: Complete recovery 10.9	13	13.6	10.5	Physical abnormalities: Complete recovery	11.8
3	Physical abnormalities: Improved 8.5	8.6	8.3	4.3	Individual function: increased	7.6
4	Physical symptoms: Relieved 8.4	6.1	4.1	0	System adjustments*	7.6
5	Attitude towards condition: Positive 6.3	5.5	8.8	9.5	Process outcomes: Care received*	6.9
6	Attitude towards M.D., care: Positive 5.8	9.6	10	14.7	Life saved	5.6

(continued)

Table 1.2 (continued)

Ranking	Beneficial outcomes					
	Sanazaro and Williamson (1970)				Our study (2017)	Total (N = 144)
	Internal medicine (N = 5554)	Surgery (N = 2804)	Paediatrics (N = 1931)	OBGYN (N = 1248)		
7	Physical symptoms: Partially relieved	5.7	0	0	0	Physical abnormalities: Prevented	4.9
8	Life saved	5.5	7.4	7.4	6.6	Unnecessary risk: Avoided or reduced	4.9
9	General improvement	5.3	0	0	0	Efficient utilization of resources*	4.9
10	Longevity: Increased	4.2	5.1	0	0	Accommodation of patient/family's needs: Positive*	4.2
11	Physical abnormalities: Prevented	0	6	6.8	5.5	Physical abnormalities: Improved	3.5
12	Compliance: Increased	0	4.2	0	0	Psychological symptoms: Partially relieved	2.8
13	Psychological symptoms: Relieved	0	0	4.1	3.8	Hospitalization: Avoided or reduced	2.8

*New subcategories under Our study (2017)

abnormalities: Complete recovery; Individual function: increased; Life saved; Physical abnormalities: Prevented; and *Physical abnormalities: Improved*). In half of the cases, the ranking was even similar (*Physical abnormalities: Complete recovery; Individual function: increased;* and *Life saved*).

Amongst the main differences, the subcategory *Attitude towards M.D., care: Positive* was the beneficial outcome most frequently reported by the managers in our sample. Furthermore, *Physical abnormalities: Improved*, which was in the third position in Sanazaro and Williamson's ranking, was not as prominent in our study, whilst *Physical abnormalities: Prevented* had a higher ranking.

In our investigation, *Unnecessary risk: Avoided or reduced, Psychological symptoms: Partially relieved* and *Hospitalization: Avoided or reduced* assumed more relevance. In Sanazaro and Williamson's (1970) work, these were not listed in the top thirteen subcategories of beneficial outcomes. Furthermore, four outcomes on our list represented a new contribution: *System adjustments; Process outcomes: Care received; Efficient utilization of resources;* and *Accommodation of patient/family needs: positive* (Amati et al. in preparation).

Detrimental Outcomes

Table 1.3 compares Sanazaro and Williamson's (1970) top thirteen most frequent detrimental outcomes with ours. Nine subcategories corresponded, five of which also had the same ranking, with similar means (i.e. *Physical abnormalities: Caused, exacerbated; Attitude towards M.D., care: Negative; Psychological symptoms: Caused, exacerbated; Physical symptoms: Caused, exacerbated;* and *Cost: Increased*). However, in our investigation *Hospitalization: Unnecessary* and *Unnecessary risk: Incurred* ranked much higher, whilst *Death caused* and *Physical abnormalities: Prolonged, unimproved* ranked lower.

Four new subcategories emerged: *Did not return to the same facility; Death not attributable to providers; Perception/Reputation of the facility: Negative;* and *Inefficient utilization of resources*. However, unlike the new process subcategories—which were at the top of the ranking—the first five most frequently reported subcategories belonged to Sanazaro and Williamson's (1970) original categorization, and three of them ranked

Table 1.3 The top thirteen most frequent detrimental outcomes reported in our study, compared to Sanazaro and Williamson (1970), expressed in percent

	Detrimental outcomes						
	Sanazaro and Williamson (1970)					Our study (2017)	
Ranking		Internal medicine (N = 3615)	Surgery (N = 2241)	Paediatrics (N = 1454)	OBGYN (N = 1056)		Total (N = 128)
1	Physical abnormalities: Caused, exacerbated	13.2	18.9	11.1	14.8	Physical abnormalities: Caused, exacerbated	13.3
2	Death caused	13	11.2	7.8	4.6	Hospitalization: Unnecessary	12.5
3	Physical abnormalities: Prolonged, unimproved	8.4	5.7	11	5.4	Unnecessary risk: Incurred	10.9
4	Attitude towards M.D., care: Negative	8.2	8.5	14.9	12.4	Attitude towards M.D., care: Negative	10.2
5	Psychological symptoms: Caused, exacerbated	7.3	4.4	6.5	8.8	Psychological symptoms: Caused, exacerbated	9.4
6	Physical symptoms: Unrelieved, prolonged	6.8	0	3.4	0	Not return to the same facility*	7.0
7	Compliance: Decreased	6.6	5.4	11.4	6.6	Death caused	5.5

(continued)

Table 1.3 (continued)

| Ranking | Detrimental outcomes | | | | Our study (2017) | Total |
| | Sanazaro and Williamson (1970) | | | | | |
	Internal medicine (N = 3615)	Surgery (N = 2241)	Paediatrics (N = 1454)	OBGYN (N = 1056)		(N = 128)
8	Physical symptoms: Caused, exacerbated	4.3	0	0	0	Physical symptoms: Caused, exacerbated	3.9
9	Cost: Increased	4.3	5.9	5	6.2	Cost: Increased	3.9
10	Hospitalization: Unnecessary	3.9	5.1	4.8	5.2	Death not attributable to providers*	3.9
11	Unnecessary risk: Incurred	0	11.1	0	9.3	Perception/Reputation of the facility: Negative*	3.9
12	Individual function: Decreased	0	3.9	0	0	Inefficient utilization of resources*	3.9
13	Attitude towards condition: Negative	0	0	5.4	5.2	Physical abnormalities: Prolonged, unimproved	2.3

*New subcategories under *Our study (2017)*

exactly as in Sanazaro and Williamson's (1970) table (*Physical abnormalities: Caused, exacerbated, Attitude towards M.D., care: Negative* and *Psychological symptoms: Caused, exacerbated*).

Discussion

Assessment is necessary for improving healthcare and the literature offers numerous examples of ways to measure quality (Griffey et al. 2015; Rushforth et al. 2015; Carinci et al. 2015). Amongst these efforts, Sanazaro and Williamson (1970) developed a classification based on physician reports of effective and ineffective performance in relation to patient outcomes. Our study replicated their design, but used a sample of US healthcare managers instead of physicians.

The findings showed that Sanazaro and Williamson's (1970) subcategories re-emerged in the episodes of care collected in this study, indicating that their suggested framework holds over time, and despite nearly fifty years of progress in quality improvement since their investigation, many issues are still relevant from the point of view of the healthcare managers. In this paper, we have presented the top fifteen effective and ineffective process subcategories and the top thirteen beneficial and detrimental outcome subcategories. In numerous cases, the ranking was quite different and new ideas were identified. There are two possible explanations for the differences: (1) contemporary healthcare managers might have different perceptions about the dimensions of quality care than do physicians; or (2) the dimensions of quality have evolved over time for both managers and physicians.

Process Subcategories

In Sanazaro and Williamson's work (1970), the most reported subcategory of effective and ineffective performance was *Arriving at diagnosis*, which emphasizes the importance attributed by physicians to identifying a condition or disease in relation to a beneficial or detrimental outcome of care. On the other hand, in our study contemporary healthcare managers seemed to identify aspects related to good teamwork (i.e. *Use of health team*) as key for the attainment of good quality care, whilst poor quality care was critically determined by poor communication amongst healthcare staff, patients and families (i.e. *Staff-patient-family communication*).

In our study, eight subcategories of effective performance and ten of ineffective performance did not even appear in the top fifteen list produced by Sanazaro and Williamson (Table 1.1). Some of them represented new contributions of our study (i.e. *Timeliness, Patient-centredness, Adherence to guidelines/protocols, Inter-staff communication* and *Staff-patient-family communication*), while others were already present in Sanazaro and Williamson's (1970) investigation but were not reported very frequently. For example, healthcare managers seemed to attribute more importance to aspects such as *Use of health team, Physician (and nurses) availability* or *Professional responsibility*, whilst they rarely discussed issues related to *Drugs, biologicals, electrolytes, fluids* or *Laboratory*.

Timeliness was not even considered as an attribution of quality by Donabedian (1990), but it later became one of the six dimensions identified by the Institute of Medicine—defined as 'reducing waits and sometimes harmful delays' (Institute of Medicine 2001, p. 40). As defined by the Agency for Healthcare Research and Quality, *timeliness* in healthcare is the 'system's capacity to provide care quickly after a need is recognized' (Agency for Healthcare Research and Quality 2016). Today, advancements in medicine and technology make it possible to intervene in and potentially solve extremely complex clinical cases; however, *timeliness* has become even more fundamental. For example, research shows that lack of *timeliness* can result in emotional distress, physical harm and higher treatment costs (Boudreau et al. 2004), whereas appropriate care delivered in a timely manner can reduce morbidity and mortality for chronic conditions such as kidney disease (Kinchen et al. 2002) and affect stroke patients' long-term disability and mortality (Kwan et al. 2004). Moreover, clinical outcomes can be improved by timely antibiotic treatments (Houck and Bratzler 2005). The relevance of *timeliness* was indeed confirmed and highlighted by our data.

Another notion that has drawn the attention of contemporary healthcare managers is *Patient-centredness*, which has been integrated into many quality definitions (Institute of Medicine 2001; World Health Organization 2006; Arah et al. 2006). There is substantial ambiguity related to its meaning and the best method to assess it (Mead and Bower 2000). We view *patient-centredness* as a partnership between the provider and the patient, and not a mere accommodation of patients' needs and expectations (Street et al. 2003). Consequently, *patient-centredness* and communication are intrinsically tied to each other: there is no *patient-centredness* without communication, but at the same time, there is no effective communication without *patient-centredness*.

Communication aspects were not absent in Sanazaro and Williamson's (1970) categorization, but they were mainly considered as part of the delivery of a service, such as instructing the patient or sending comfort messages, and not as an interplay amongst all parties involved. Contemporary research attributes to provider—patient communication historical functions such as exchanging information or responding to patients' emotions, but it also sheds light on additional ones, such as fostering healing relationships, managing uncertainty, making decisions with the active involvement of patients and families, and enabling patients' self-management whilst advocating for patients and supporting their autonomy (Epstein and Street 2007). In this investigation, communication—with its different facets—accounted for almost 50% of both effective and ineffective performance, confirming the growing awareness of its importance in healthcare (Agarwal et al. 2010).

Finally, the emergence of the subcategory *Adherence to guidelines/protocols* suggests that it is an increasingly important topic, as it has been shown that in the USA only 55% of patients receive care as recommended in the guidelines (McGlynn et al. 2003). Research studies are trying to uncover the barriers that hinder the implementation of guidelines in clinical practice (Lugtenberg et al. 2011).

Outcome Subcategories

Concerning beneficial outcomes of care, in both investigations the second most discussed beneficial subcategory was *Physical abnormalities: Complete recovery*. However, in Sanazaro and Williamson's (1970) work, the first one was *Individual function: increased*, whilst in the episodes provided by our participants it was *Attitude towards medical doctors and care*. This denotes contemporary healthcare managers' awareness and concern that the quality of care affects more than physical and psychological patient outcomes. In fact, amongst the new beneficial outcome subcategories we found *Accommodation of patient/family needs*, whilst amongst the new detrimental ones we found *Not return to the same facility* and *Perception/Reputation of the facility*.

In both studies, the most frequently reported detrimental outcome was *Physical abnormalities: Caused, exacerbated*. Contemporary healthcare managers are concerned—as were physicians fifty years ago—that the care provided may not improve patient health, but instead it may prompt or worsen physical abnormalities, diseases, conditions and their complications. Surprisingly, despite the increasing attempts to contain healthcare costs (Schnipper et al. 2012; Minogue and Wells 2016), there was no qualitative difference in the ranking of *Cost: Increased*. Whilst we typically expect managers to factor in costs in their assessment of quality of care, the respondents in our study did not emphasize financial aspects very much.

On the other hand, the importance of Sanazaro and Williamson's (1970) subcategories *Unnecessary risk* and *Hospitalization* are perfectly in line with current management concerns. This was also emphasized by the emergence of new subcategories such as *System adjustments*, *Utilization of resources* and *Perception/Reputation of the facility*. In fact, much research has been conducted to investigate and address issues such as rehospitalization (Hansen et al. 2013), misuse of resources (Bulger et al. 2013), or hospital reputation (Mira et al. 2013).

Limitations

The limitations of our investigation mostly pertain to sample size and that it included alumni from only one US graduate programme in Healthcare Administration, who mainly work in the same geographic area in which they earned their degree. The response rate was 33%, which limits the validity of the results, even though it is similar to that achieved by other surveys of healthcare managers (McDonagh and Umbdenstock 2006; Vaughn et al. 2014). Finally, we compared contemporary managers with physicians. Different stakeholders account for diverse perspectives and findings. For this reason, further research is needed to focus on contemporary physicians in order to investigate the evolution of quality dimensions in relation to this specific group of stakeholders.

Conclusion

This study replicated Sanazaro and Williamson's (1970) design to investigate qualitatively how the dimensions of quality have evolved over time and how the perceptions of managers might be different from those of physicians. Our findings confirmed the existence of the subcategories identified about fifty years ago by Sanazaro and Williamson (1970) in relation to the process and outcomes of care, suggesting that those dimensions of quality are still valid nowadays. However, several subcategories gained more importance, and new dimensions emerged from the data. This suggests that the multifaceted concept of quality care has evolved over time, and for this reason, it is imperative to take into account a wide spectrum of dimensions when assessing it, and to potentially change priorities in the process of continuous quality improvement.

References

Agarwal, R., Sands, D. Z., & Schneider, J. D. (2010). Quantifying the economic impact of communication inefficiencies in US hospitals. *Journal of Healthcare Management, 55*(4), 265–282.

Agency for Healthcare Research and Quality. (2016). *Chartbook on access to health care: Timeliness.* Retrieved from http://www.ahrq.gov/research/findings/nhqrdr/2014chartbooks/access/access-time.html. Accessed 2 Dec 2016.

Amati, R., Kaissi, A. A., Brook, R. H., & Hannawa, A. F. (in preparation). *U.S. health care managers' perceptions of good and poor quality: A grounded taxonomy of quality based on evidence from episodes of care.*

Arah, O. A., Westert, G. P., Hurst, J., & Klazinga, N. S. (2006). A conceptual framework for the OECD health care quality indicators project. *International Journal for Quality in Health Care, 1*, 5–13.

Boudreau, R. M., McNally, C., Rensing, E. M., & Campbell, M. K. (2004). Improving the timeliness of written patient notification of mammography results by mammography centers. *The Breast Journal, 10*(1), 10–19.

Brook, R. H., McGlynn, E. A., & Cleary, P. (1996). Measuring quality of care: Part 2: Measuring quality of care. *The New England Journal of Medicine, 335*(13), 966–970.

Bulger, J., Nickel, W., Messler, J., Goldstein, J., O'Callaghan, J., Auron, M., & Gulati, M. (2013). Choosing wisely in adult hospital medicine: five opportunities for improved healthcare value. *Journal of Hospital Medicine, 8*(9), 486–492.

Campbell, S. M., Roland, M. O., & Buetow, S. A. (2000). Defining quality of care. *Social Science and Medicine, 51*, 1611–1625.

Carinci, F., Van Gool, K., Mainz, J., Veillard, J., Pichora, E. C., Januel, J. M., & Klazinga, N. S. (2015). Towards actionable international comparisons of health system performance: expert revision of the OECD framework and quality indicators. *International Journal for Quality in Health Care, 27*(2), 137–146.

Donabedian, A. (1966). Evaluating the quality of medical care. *The Milbank Memorial Fund Quarterly, 44*(3), 166–203.

Donabedian, A. (1988). The quality of care: How can it be assessed? *JAMA, 260*(12), 1743–1748.

Donabedian, A. (1990). The seven pillars of quality. *Archives of Pathology and Laboratory Medicine, 114*(11), 1115–1118.

Donabedian, A. (1996). Evaluating the quality of medical care. *The Milbank Memorial Fund Quarterly, 44*(3), 166–203.

Donabedian, A. (1998). The quality of care: How can it be assessed? *JAMA, 260*(12), 1743–1748.

Epstein, R. M., & Street, Jr. R. L. (2007). *Patient-Centered Communication in Cancer Care: Promoting healing and reducing suffering.* National Cancer Institute, NIH Publication No. 07-6225.

Griffey, R. T., Pines, J. M., Farley, H. L., Phelan, M. P., Beach, C., Schuur, J. D., & Venkatesh, A. K. (2015). Chief complaint-based performance measures: a new focus for acute care quality measurement. *Annals of Emergency Medicine, 65*(4), 387–395.

Hansen, L. O., Greenwald, J. L., Budnitz, T., Howell, E., Halasyamani, L., Maynard, G., & Williams, M. V. (2013). Project BOOST: effectiveness of a multihospital effort to reduce rehospitalization. *Journal of Hospital Medicine, 8*(8), 421–427.

Houck, P. M., & Bratzler, D. W. (2005). Administration of first hospital antibiotics for community-acquired pneumonia: Does timeliness affect outcomes? *Current Opinion in Infectious Diseases, 18*(2), 151–156.

Institute of Medicine. (1999). *To err is human: Building a safer health system.* Washington, DC: National Academy Press.

Institute of Medicine. (2001). *Crossing the quality chasm: A new health system for the 21st century.* Washington, DC: National Academy Press.

Kinchen, K. S., Sadler, J., Fink, N., Brookmeyer, R., Klag, M. J., Levey, A. S., & Powe, N. R. (2002). The timing of specialist evaluation in chronic kidney disease and mortality. *Annals of Internal Medicine, 137*(6), 479–486.

Kwan, J., Hand, P., & Sandercock, P. (2004). Improving the efficiency of delivery of thrombolysis for acute stroke: A systematic review. *Quarterly Journal of Medicine, 97*(5), 273–279.

Lugtenberg, M., Burgers, J. S., Besters, C. F., Han, D., & Westert, G. P. (2011). Perceived barriers to guideline adherence: A survey among general practitioners. *BMC Family Practice, 12*(1), 98.

Maguard, A. (2006). The modern quality movement: Origins, development and trends. *Total Quality Management, 17*(2), 179–203.

McGlynn, E. A., Asch, S. M., Adams, J., Keesey, J., Hicks, J., DeCristofaro, A., & Kerr, E. A. (2003). The quality of health care delivered to adults in the United States. *New England Journal of Medicine, 348*(26), 2635–2645.

McDonagh, K. J., & Umbdenstock, R. J. (2006). Hospital governing boards: A study of their effectiveness in relation to organizational performance. *Journal of Healthcare Management, 51*(6), 377.

Mead, N., & Bower, P. (2000). Patient-centeredness: A conceptual framework and review of the empirical literature. *Social Science and Medicine, 51,* 1087–1110.

Minogue, V., & Wells, B. (2016). Reducing waste in the NHS: An overview of the literature and challenges for the nursing profession. *Nursing Management, 23*(4), 20–25.

Mira, J. J., Lorenzo, S., & Navarro, I. (2013). Hospital reputation and perceptions of patient safety. *Medical Principles and Practice, 23*(1), 92–94.

Rushforth, B., Stokes, T., Andrews, E., Willis, T. A., McEachan, R., Faulkner, S., & Foy, R. (2015). Developing 'high impact' guideline-based quality indicators for UK primary care: a multi-stage consensus process. *BMC Family Practice, 16*(1), 1.

Sanazaro, P. J., & Williamson, J. W. (1970). Physician performance and its effects on patients: A classification based on reports by internists, surgeons, pediatricians, and obstetricians. *Medical Care, 8*(4), 299–308.

Schnipper, L. E., Smith, T. J., Raghavan, D., Blayney, D. W., Ganz, P. A., Mulvey, T. M., & Wollins, D. S. (2012). American Society of Clinical Oncology identifies five key opportunities to improve care and reduce costs: the top five list for oncology. *Journal of Clinical Oncology, 30*(14), 1715–1724.

Street, R. L., Jr., Krupat, E., Bell, R. A., Kravitz, R. L., & Haidet, P. (2003). Beliefs about control in the physician–patient relationship: Effect on communication in medical encounters. *Journal of Internal Medicine, 18*, 609–616.

Vaughn, T., Koepke, M., Levey, S., Kroch, E., Hatcher, C., Tompkins, C., & Baloh, J. (2014). Governing Board, C-suite, and Clinical Management Perceptions of Quality and Safety Structures, Processes, and Priorities in US Hospitals. *Journal of Healthcare Management, 59*(2), 111–129.

World Health Organization. (2006). *Quality of care.* Geneva: World Health Organization.

2

Multi-level Pluralism: A Pragmatic Approach to Choosing Change and Improvement Methods

Liz Wiggins and Brian Marshall

Introduction

The need for healthcare organizations and systems to improve and sustain quality is uncontentious. In most developed economies, projected health spend is outstripping GDP growth as a result of significantly changing demographics, advances in medicine, surgical techniques and patient expectations. Politicians, the media, professionals and patients all have views as to how the quality of patient care can be improved whilst spending is reduced. Whatever the latest government white paper, and whether framed as modernization (Freeman and Peck 2010), culture change (Braithewaite et al. 2010) or quality improvement (Berwick 2009), leaders are needed who have the skills

L. Wiggins (✉) · B. Marshall
Ashridge Executive Education at Hult International Business School,
Berkhamsted, Hertfordshire, UK
e-mail: liz.wiggins@ashridge.hult.edu

and capability to translate those visions into reality on the wards, in the GP surgery, in the recovery college. Leading the improvement of quality in healthcare is arguably, therefore, one of the most challenging areas of modern leadership (Gregory et al. 2012).

The array of approaches to organizational change and improvement is vast (Langley et al. 2009; Myers et al. 2012). There is an extensive body of knowledge termed the 'improvement sciences' (Shewhart 1931; Deming 1986; Goldratt and Cox 2004; Womack and Jones 2003), meaning ways of thinking about improvement which are evidence-based and often involve analysis of quantitative data. In the field of leadership and change, leaders are offered linear change approaches (Kotter 1995) or the identification of adaptive and whole system challenges (Heifetz 2002), through to the emergent change of Stacey (2010, 2012a, b) and Shaw (2002). These approaches are underpinned by different ontologies from modernism to post-positivism, through systems thinking and into complexity, making it difficult to answer questions about which is best for any given situation.

For change to be sustainable, leaders arguably need to consider people, paying attention to staff, patients and carers. To understand how best to relate to people and intervene in group dynamics, there are numerous psychological theories such as Transaction Analysis and Gestalt (Lapworth and Sills 2011), and from organizational development, theories such as dialogue (Isaacs 1999) and Appreciative Inquiry (Cooperrider and Whitney 2005).

Faced with such an overwhelming choice of approaches to change and improvement, the leader may well feel daunted, believing that '[c]hange is like a totem before which we must prostrate ourselves and in the face of which we are powerless' (Grey 2005, p. 90). There are thus a number of dilemmas for health leaders: how do they choose what change or improvement approach to use in a particular situation? Does pick and mix work, or will that just confuse everyone? Is it better to choose one approach and stick to it?

We propose a typology of reactions to these dilemmas, which is explored in this chapter. Our typology includes Singularism, Conflation, Privileging, Unaware Pluralism and Multi-level Pluralism.

Our Interest in This and Our Methods

This paper, and the thinking behind it, emerged from a leadership development programme at Ashridge Business School, designed and delivered by the authors. This programme, marketed as GenerationQ but known academically as the Ashridge Masters in Leadership (Quality Improvement), is designed for senior clinical, managerial and policy leaders in healthcare in the UK. It seeks to equip them to lead the improvement of healthcare delivery in their highly challenging context.

This Masters level programme has, from the beginning, been informed by different perspectives about how to effect change in healthcare organizations, embracing as it does both technical quality improvement disciplines, such as Lean, Theory of Constraints and Six Sigma, as well as more relational approaches from Organizational Development.

In endeavouring to make sense of the different theories and approaches available and the participants' responses to them, we have been exploring the notion of pluralism as a potentially useful framing of some apparent clashes in ontology and methodology.

Our method has been to devise this framework based on our own observations and reflective practice, and then to engage in Action Research with a broad cross section of our programme participants. Reason and Bradbury state that '[a] primary purpose of action research is to produce practical knowledge that is useful to people in the everyday conduct of their lives' (2001, p. 2). Whilst recognizing that Action Research is an orientation to research rather than a specific methodology (Ladkin 2007), this emphasis on what is useful felt appropriate given our interest in the practical dilemmas faced by leaders.

We have therefore engaged in cycles of first and second person enquiry with almost one hundred past participants, as individuals and in group sessions, inviting them to be co-researchers.

Defining Pluralism

The metaphysical aspects of pluralism, and whether or not a pluralist ontology is tenable, have been explored and staunchly defended in philosophical circles (McDaniel 2009; Turner 2010). The latter argues that only a pluralist view can reflect the complexity of reality, offering a 'metaphysically perspicuous' approach (ibid., p. 8). In the field of organizational research, some writers have sought to find a route that recognizes the strengths of modernist and post-modern research and enquiry methods, since 'a single paradigm is necessarily limiting' (Lewis and Kelemen 2002, p. 252). Modernism embraces beliefs about reason and progress, and from this network of beliefs chooses (either consciously or otherwise) to focus on and privilege certain voices and views whilst playing down others, especially those which reflect ambiguity and uncertainty. Post-modern research, on the other hand, seeks to emphasize the uncertainty of organizational life and to find an approach which is congruent with this by stressing fragmented pieces of information and offering a patchwork quilt of impressions of the subject matter.

Multi-paradigm enquiry potentially offers a new look at this modern versus post-modern duality. Whereas use of a single paradigm can produce a valuable but narrow view, multi-paradigm enquiry may foster 'more comprehensive portraits of complex organisational phenomena' (Gioia and Pitre 1990, p. 587). Lewis and Keleman (2002, p. 258) explain this further:

> Multi-paradigm researchers apply an <u>accommodating</u> ideology, valuing paradigm perspectives for their potential to inform each other toward more encompassing theories.

It is in this area of multiple perspectives, of 'both … and', that our recent work in leadership development has focused. We are becoming increasingly convinced that a pluralist approach to change and improvement holds exciting new ways of approaching some of today's toughest leadership challenges and provides a potential answer to the dilemmas for health leaders posed earlier in this chapter.

Revealing Underpinning Assumptions in Three Change Approaches

In this section, we take Lean, Appreciative Inquiry and Complex Social Processes as three different approaches to change in complex systems and reveal their underpinning and sometimes contradictory assumptions, acknowledging that some subtleties will be lost in summarizing. Their fundamental differences serve as a good illustration of our central proposition.

Lean

Originating with figures such as Walter Shewhart and Edwards Deming, Lean came to fruition in the Toyota Production System. Womack and Jones (2003) identify five core principles of Lean Thinking:

i. Specify the value as desired and judged by the customer or end user.
ii. Identify 'value streams' (the process from end to end) for each product or service providing that value and identify and systematically remove any waste.
iii. Make the product or service flow continuously.
iv. Introduce pull (meaning only move goods where there is demand further down the value chain) between all steps where continuous flow is impossible.
v. Strive for perfection through continuous improvement for each value stream.

Here, the invitation is to see organizations as existing to satisfy and exceed customer demands; organizations are collections of 'value streams'. If those value streams do nothing but add value and eliminate waste, we have a long-term prescription for sustainable high-quality organizations.

Appreciative Inquiry

Appreciative Inquiry (AI) originated in Case Western University (Cooperrider and Whitney 2005; Barrett and Fry 2005). The underlying philosophy of AI is relatively explicit, relying on both social constructionism and the 'heliotropic hypothesis'.

Social constructionism (Weick 1995) suggests that social reality is a construction agreed upon by the members of that society. Thus organizational reality is only bounded by our collective imaginations and by our ability to envision a different future. Creating new and better ideas, and using new and different language, is, therefore, a powerful way of changing organizations. The heliotropic hypothesis suggests that organizations and social systems evolve towards the most positive image they hold of themselves. Both these underpinning theories, therefore, suggest that by finding ways of helping people think and dream together more positively, there will be natural movement towards that improved state.

Complex Social (or Responsive) Processes

Stacey (2010, 2012a, b), postulates that thinking about organizations as spatial entities which exist apart from the people who populate them is unhelpful. He suggests that organizing is a constantly iterated process of gesture and response between people. Meaning arises in those interactions in every moment. As organizing is a complex (in the sense of the Complexity Sciences) process, no one (including leaders) can predict or control the direction the organization will take—even though they may be given ostensible responsibility by others. They may be in charge, but not in control (Stacey 2010, p. 233).

In terms of organizational change, this theory emphasizes the following:

i.   Change takes place in conversation and everyday interactions not in the grand announcement or change programme.
ii. Change emerges as people interact together.
iii. The leader's role is to judge when to hold a conversation open and to notice and amplify emerging patterns.

Of the three approaches considered here, a Complex Responsive Process (CRP) view of organizing has the least to say as a method of organizational change, precisely because it seeks to shed light on organizing rather than offering a prescription for change. However, Rodgers (2006) and Shaw (2002) both offer the possibility of generative change through taking a CRP view.

Our Emerging Proposition of Multi-level Pluralism

Our contention is that a leader in healthcare, attempting to improve quality and patient outcomes, faces what can be categorized as wicked (Grint 2008) and complex (McCandless 2008) problems. They will thus need to employ a range of improvement and change methods, but their dilemma will be which to choose. This is problematic as these approaches clash at different levels, as shown in Table 2.1 below. Our proposition is that rather than requiring a 'numbing' thought process, by finding ways to reconcile, integrate or conflate the different approaches, multi-level pluralism is not only possible but may also help to unlock the full power of each approach. By pluralism, we mean adopting an approach in which two or more states, groups or principles can coexist. We suggest that this can be at a number of levels including ontology, ideology and methodology; hence the approach is multi-level.

Table 2.1 Comparison of approaches

	Lean	Appreciative inquiry	Complex responsive processes
Ontology	Modernist Knowable reality Positivism	Post-modernist Reality is socially constructed	
Epistemology	Empirical data is knowledge	Meaning constantly shifts—eclectic approaches to knowing	
Ideology (of change)	Change must be structured Consistent leadership to encourage widespread use	Organizations grow naturally towards the sun	Change is always happening—no one can be said to be in control
Methodology	Measurement, analysis, improvement, control to eliminate waste	Choreographed appreciative story-telling, amplified to encourage change	Conversations are building blocks of change

To fully utilize these approaches, the leader is knowingly or unknowingly embracing a linked set of attendant assumptions and views. For example, a leader advocating improvement through using Lean methodology is (perhaps unwittingly) also acting from a positivist, empirically based world view. A leader advocating AI is acting from a social constructionist ontology.

So, how can an individual who believes wholeheartedly in the efficacy of the Lean approach, with its emphasis on control and the elimination of variation, see the merit in Complex Social Processes where the leader cannot be said to be in control, and where variation is seen as a rich source of newness and innovation? How can someone who believes that positive psychology and appreciative thinking naturally encourage organizational movement feel comfortable with a Lean approach, which seeks to surface problems and deficits? If operating from one paradigm or world view, it can be hard to see merits in another, as Kuhn (2012) describes in his history of scientific revolutions.

Potential Responses

In our work as leadership developers, working alongside clinical, managerial and policy leaders, we have seen various ways of dealing with the conflict between different change and improvement approaches. We summarize this into five ways of thinking about the issue:

i.   Singularism.
ii. Conflation.
iii. Privileging.
iv. Unaware Pluralism.
v.   Multi-level Pluralism.

We explore these different responses below, recognizing that our typology is an analytically convenient way of categorizing different responses to embedded pluralistic assumptions. We also note that in our work with leaders, individuals can be ontologically flirtatious, flitting between combinations of different responses at different times.

Singularism

Often practitioners of a single approach advocate their position with an almost religious fervour, as the way. This espoused certainty remains a common feature in change initiatives, perhaps because it is congruent with the visionary, heroic styles of leadership frequently found in health-care settings (Binney et al. 2005). Singularism seems to be the default position for participants beginning our Masters leadership development programme. Despite knowing that their context is complex and political, they frequently start with the assumption (or hope) that there will be a single methodology, a silver bullet for all of their organizational change needs. Early excitement and short-term gains often lead to disillusion-ment or challenges in sustaining or embedding a specific approach.

Conflation

Perhaps equally as frequent is the tendency to conflate different approaches, reducing them to their lowest denominators. Phrases such as 'Really this is just a matter of common sense' or 'Implementing Lean is bound to be complex' seek to reconcile different approaches to organizational change into some kind of homogenous whole. However, to achieve some form of harmonious reconciliation the sharp edges of each approach must be removed, their differences lost.

To illustrate why this is simply unsound and a dumbing down of the theory, consider the contrast in thinking between AI and Lean, shown in Table 2.2.

These differences at a theoretical level lead to fundamentally differ-ent ways of approaching organizational issues in practice—amplifying or dampening difference, for example, or searching for problems ver-sus paying attention to strengths. Conflating the two approaches into

Table 2.2 Contrasting thinking between Lean and AI

Lean	Appreciative inquiry
It is possible (and desirable) to reduce variation and thus create greater efficiency	Differences in perspectives and ways of doing things are inevitable and welcome. Variation leads to positive change
No problem is a problem—only by surfacing what is going wrong can we fix it	Focus on what is already working, the best of what is. Deficit-based thinking does not take us forward

one is simply not possible without losing the internal integrity of each approach.

Privileging

An alternative temptation is some form of privileging. Whilst perhaps more logically sound than conflation, this risks raising or lowering the adjudged worth of particular theoretical approaches. For example, it may be tempting to see organizational issues exclusively through the lens of Complex Social Processes, using Stacey's grid which he later rejected (2010, 2012a) (see Fig. 2.1).

Stacey suggests that organizations need both stability and instability at the same time. The temptation may be to try and 'locate' other theories within the grid. Perhaps Lean fits in the bottom left-hand corner, with AI more in the emergent space further out? We advocate caution here because of the hierarchy which this kind of thinking suggests. Believing Lean fits within an overall framework of Complex Social Processes relegates Lean to a limited view of the world which only applies in certain circumstances, and similarly with AI. Thinking this way promotes Complex Social Processes to the top slot, to being the single unifying framework which encapsulates the other two. Rather than adopting a pluralistic approach, one has been chosen over the other two.

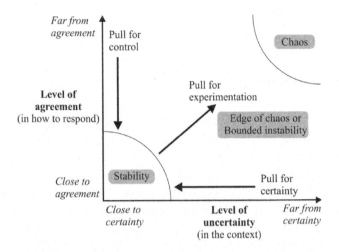

Fig. 2.1 Stacey's grid of complex social processes

The risk of privileging is that it may prevent people from fully utilising the depth of different approaches.

Unaware Pluralism

We know from working with healthcare leaders that they prefer pragmatic solutions, often manifesting an inbuilt caution around anything that sounds too theoretical and impractical. It is perfectly possible, and sometimes effective, to have an eclectic approach, a sort of bricolage—a kind of unaware pluralism which enables flexibility and context-appropriate approaches without ever unearthing the theoretical underpinnings. We are not advocating that all leaders need to fully explore the rarefied aspects of ontology and epistemology, but we do believe that some exploration of these areas brings benefit. If they are unaware of the underlying fundamentals of change methodologies, leaders risk being surprised when an approach to which they are wedded as the 'truth' is rejected by some, or when a method is not as powerful as anticipated or change is hard to sustain.

Multi-level Pluralism

In advocating multi-level pluralism in response to the change challenges faced by healthcare leaders, we suggest that we have the capacity as human beings to hold a pluralist view when it comes to matters as complex as organizational change—that we are capable of believing each of these approaches is valid as one perspective on how organizations work and change may come about, and only by holding and using all of them do we get the fullest possible range of understanding and action to cope with the complexity and challenge of modern organizational life, especially in healthcare.

This differs from an ecumenical or simply tolerant view, in that at any one time we may fully and wholeheartedly subscribe to the world view which underpins each of these theories. We authentically believe that an organization can be a set of value-adding processes or streams (Lean) and that organizing is a constantly iterated dance of gesture and response (CRP).

When these views collide, as we believe they will, we are suggesting leaders need to live with the dilemmas, paradoxes and ambiguities that emerge. This has parallels with the debate in quantum physics about whether light consists of particles or waves. Is this duality paradoxical or do wave-particle aspects always coexist (the de Broglie Bohm theory)? Niels Bohr (Kumar 2011) regarded the 'duality paradox' as a fundamental or metaphysical fact of nature. Others have refuted such thinking, insisting that light is made of particles which sometimes behave like waves. We are drawn to Einstein's words on this:

> It seems as though we must use sometimes the one theory and sometimes the other, while at times we may use either. We are faced with a new kind of difficulty. We have two contradictory pictures of reality; separately neither of them fully explains the phenomena of light, but together they do'. (quoted in Harrison 2002)

Similarly, we believe that to understand organization improvement, contradictory 'pictures of reality' must be embraced. Leaders, faced with the dilemma of which improvement approach to adopt, need to hold multiple perspectives on how organizations are and how they change, even if these perspectives present fundamentally different ontologies. In short, they need to be pluralist.

To illustrate further how this pluralism operates at multiple levels, the examples summarized in Fig. 2.1 all differ at a methodological level. Whilst Lean differs from both Appreciative Inquiry and Complex Responsive Processes at an ontological level, Appreciative Inquiry and Complex Responsive Processes share a post-modern ontology. However, when considering what we have termed their ideology of change, by which we mean what is valued in effecting organizational change, the two theories diverge. Appreciative Inquiry holds that focusing on positive conversations is the route to success, whilst Complex Responsive Processes suggests this is unhealthy and unrealistic. Thus the pluralist leader may have to embrace differences and paradoxes at different levels.

Testing Out with Health Leaders

Our thinking about multi-level pluralism arose from working with healthcare leaders who were also participants on a leadership development programme. It was therefore with them that we tested our emerging proposition, drawing on the principles of Action Research.

In this section, we lightly draw attention to three emerging themes from this enquiry which both validate the usefulness of the idea of multi-level pluralism and raise questions for further research and practice.

The first theme is that of relief. Many spoke of the way the idea of multi-level pluralism helped them make sense of, and validate, their own personal responses to the differences between improvement and change approaches to which they had been exposed. Typical comments were: 'It frames what I feel'; 'It is incredibly helpful'; or 'It makes sense of what it is we have been learning and the differences I see in my organization'. One consultant described his emerging pluralism in this way:

> I have gone from wearing one hat all the time to having many different hats and choosing which one which is the most appropriate in the context in which I find myself. ... I still make the odd fashion faux pas but thankfully less often.

Such comments offer initial validation of the usefulness of multi-level pluralism as a means to make sense of, and work with, different change and improvement approaches.

Second, there is a general welcoming of the framework itself and the typology. Some drew attention to the dangers of a singularist approach, noting: 'It has the potential to cause elitism ... and can result in ... marginalising the "zealots with their strange language", resulting in counterproductive behaviours amongst staff'. Others found that explicitly identifying conflation as a potential response helped them to recognize a pattern in their own behaviour. 'A learning point for me has been how to avoid the temptation of plucking the best bits from the theories and creating a Frankenstein monster of QI techniques.'

Third, questions of a practical nature were raised, such as: how and when could multi-level pluralism be usefully introduced to leaders? What might be the impact on the followers, and indeed the bosses, of a leader who embraces pluralism? Would a pluralist be seen by others as being inauthentic, indecisive or 'flip flopping'? Would providing a 'voice over' to explain the different choices being made mitigate this?

Conclusion and Further Considerations

We began this chapter by suggesting that health leaders face a dilemma when confronted by the vast array of change and improvement methods. We propose that multi-level pluralism may be a route for making sense of different approaches by drawing attention to the underpinning ontological, epistemological, ideological and methodological differences. Initial validation with leaders suggests this is the case.

Further exploration is required into the practical use and introduction of multi-level pluralism. However, we believe that the concept gives leaders increased confidence that they can deal with the multiple change challenges they face at work, and means they will be less susceptible to the guile of quick fixes or the certainty of a promised right way. Given the importance of improving patient care and delivering a high level of service at an affordable cost, we can think of few other areas where the stakes and potential rewards are so high—not just for healthcare leaders but for all of us.

References

Barrett, F. J., & Fry, R. E. (2005). *Appreciative inquiry: A positive approach to building cooperative capacity*. Chagrin Falls: Taos Institute.

Berwick, D. (2009). What patient centred should mean: Confessions of an extremist. *Health Affairs, 28*(4), 555–565.

Binney, G., Wiles, G., & Williams, C. (2005). *Living leadership: A practical guide for everyday heroes*. Harlow: FT/Prentice Hall.

Braithewaite, J., Hyde, P., & Pope, C. (Eds.). (2010). *Culture and climate in health care organizations*. Basingstoke: Palgrave.

Cooperrider, D., & Whitney, D. (2005). *Appreciative inquiry: A positive revolution in change*. New York: Berrett-Koehler.

Deming, W. E. (1986). *Out of the crisis*. Boston, MA: MIT Press.

Freeman, T., & Peck, E. (2010). Culture made flesh: Discourse, performativity and materiality. In J. Braithewaite, P. Hyde, & C. Pope (Eds.), *Culture and climate in health care organizations*. Basingstoke: Palgrave.

Gioia, D. A., & Pitre, E. (1990). Multiparadigm perspectives on theory building. *Academy of Management Review, 15*, 584–602.

Goldratt, E., & Cox, J. (2004). *The goal: A process of ongoing improvement*. Great Barrington, MA: North River Press.

Gregory, S., Ham, C., & Dixon, A. (2012). *Health policy under the coalition government—A mid-term assessment*. Retrieved from http://www.kingsfund. org.uk/publications/health-policy-under-coalition-government. Accessed May 2 2016.

Grey, C. (2005). *A very short, fairly interesting and reasonably cheap book about studying organizations*. Sage.

Grint, K. (2008). Wicked problems and clumsy solutions: The role of lead ship. In *Clinical Leader*, *1*(2), pp. 12–15. Stockport: BAMM.

Harrison, D. (2002). *Complementarity and the Copenhagen interpre tion of quantum mechanics. UPSCALE.* Toronto: Department of Physi University of Toronto.

Heifitz, R. (2002). *Leadership on the line: Staying alive through the dangers leading.* Boston, MA: Harvard Business School Press.

Isaacs, W. (1999). *Dialogue and the art of thinking together.* New York: Banta Doubleday Dell Publishing Group.

Kotter, J. P. (1995). Leading change, Why transformation efforts fail. *Harva Business Review, 73*, 59–67.

Kuhn, T. (2012). *The structure of scientific revolutions* (4th ed.). Chicag University of Chicago Press.

Kumar, M. (2011). *Quantum: Einstein, Bohr, and the great debate about t nature of reality* (Reprint ed.). New York: W. W. Norton & Company.

Ladkin, D. (2007). Action research. In C. Seale, G. Gobo, J. Gubrium, & l Silverman (Eds.), *Qualitative research practice* (pp. 478–490). London: Sag

Langley, G. J., Moen, R., Nolan, K., Nolan, T. W., Norman, C. L., & Provo L. P. (2009). *The improvement guide: A practical approach to enhancing orga izational performance.* London: Wiley.

Lapworth, P., & Sills, C. (2011). *An introduction to transactional analys Helping people change.* London: Sage.

Lewis, M. W., & Kelemen, M. L. (2002). Multiparadigm inquiry: Explori organizational pluralism and paradox. *Human Relations, 55*(2), 251–275.

McCandless, K. (2008). *Safely taking risks: Complexity and patient safe Seattle, WA: Social Invention Group.

McDaniel, K. (2009). Ways of being. In D. Chalmers, D. Manley, & Wasserman (Eds.), *Metametaphysics: New essays on the foundations of ont ogy.* Oxford: Oxford University Press.

Myers, P., Hulks, S., & Wiggins, L. (2012). *Organizational change: Perspectiv on theory and practice.* Oxford: Oxford University Press.

Reason, P., & Bradbury, H. (Eds.). (2001). *Handbook of action researc Participative inquiry.* London: Sage.

Rodgers, C. (2006). *Informal coalitions: Mastering the hidden dynamics of orga izational change.* London: Palgrave Macmillan.

Shaw, P. (2002). *Changing conversations in organizations: A complexity approa to change (Complexity and emergence in organizations).* London: Routledge.

Shewhart, W. A. (1931). *Economic control of quality of manufactured product.* New York: Van Nostrand.

Stacey, R. (2010). *Strategic management and organizational dynamics: The challenge of complexity* (6th ed.). London: Financial Times/Prentice Hall.

Stacey, R. (2012a). *The tools and techniques of leadership and management: Meeting the challenge of complexity.* London: Routledge.

Stacey, R. (2012b). *The paradox of consensus and conflict in organizational life.* Retrieved from https://complexityandmanagement.wordpress. com/2012/09/22/287/. Accessed June 11 2015.

Turner, J. (2010). Ontological pluralism. *Journal of Philosophy, 107*(1), 5–34.

Weick, K. (1995). *Sensemaking in organizations.* Thousand Oaks, CA: Sage.

Womack, J., & Jones, D. (2003). *Lean thinking: Banish waste and create wealth in your corporation.* London: Simon & Schuster/Free Press.

3

Amendments to Reporting of QI Interventions: Insights from the Concept of Affordances

Emilie Berard, Jean-Louis Denis, Olivier Saulpic and
Philippe Zarlowski

Introduction

Quality improvement (QI) interventions, a managerial technology, are used extensively in healthcare teams and organizations to solve problems associated with the quality, effectiveness and efficiency of care. However, studies of the outcomes of particular QI interventions have failed to demonstrate consistent positive effects across healthcare settings (e.g. Grimshaw et al. 2004; Schouten et al. 2008).

E. Berard (✉)
ITESO, San Pedro Tlaquepaque, Jalisco, Mexico
e-mail: berard@iteso.mx

J.-L. Denis
Université de Montréal, Québec, Canada

O. Saulpic · P. Zarlowski
Management Control Department, ESCP Europe, Paris, France

To increase the value of research in this area, a growing number of studies is focusing on contextual factors that might explain differences in outcomes. It is now accepted that 'context matters' (Tomoaia-Cotisel et al. 2013) to the effectiveness of an intervention, especially in complex organizational and work environments, where the control required for randomization processes typical of clinical trials is difficult (Stevens and Shojania 2011). In order to enhance the replicability of outcomes with a given intervention, it has been proposed that research into QI interventions should concentrate on identifying contextual factors that are likely to cause outcome variations (Shojania 2013).

A first generation of healthcare studies, based on in-depth interviews and expert assessment, identified the main drivers of context, as well as their underlying relation to QI design (Kaplan et al. 2010; Dixon-Woods et al. 2011; Øvretveit 2011; Bate et al. 2014). Other studies, aiming to improve the internal and external validity of QI research, have developed comprehensive, multilevel and multidimensional frameworks to systemically identify, categorize and report on contextual factors and the interactions between intervention and context (Kaplan et al. 2012; Tomoaia-Cotisel et al. 2013; Kringos et al. 2015; Ogrinc et al. 2015). Overall, this stream of research in the QI literature focuses on the relationship between QI design and the implementation context with the aim of informing changes in the healthcare working environment.

Interestingly, this concern mirrors recent developments in organization and management studies. In studying organizational change processes, various authors are interested in the capacities for action that new techniques either create or constrain, considering that it is the interplay between techniques and social processes that account for changes in practice. Drawing on the concept of affordances, which refers to the possibilities for action offered by the technical and interpretive properties of an object (Hutchby 2001), they aim to improve understanding of the complex dynamics that accompany the introduction of new managerial technologies in organizational settings (Jarzabkowski and Kaplan 2014; Orlikowski and Scott 2008).

Whilst these two streams of QI and organization studies arise from distinct traditions in different fields, they nonetheless share a concern for how changes in practice relate to the interplay between actors and techniques that are new to them. Given this common central objective, we argue that integrating knowledge from recent studies around affordances into QI literature might provide a complementary understanding of recent results and methodological developments.

This chapter first presents two recent developments in the QI literature that enable comprehensive and systematic study of interventions and their context of implementation. MUSIQ (Kaplan et al. 2012) captures key contextual dimensions for the study of implementation dynamics, and SQUIRE (Ogrinc et al. 2008; Goodman et al. 2016) provides a framework for the systematic reporting of QI interventions implementation. Both models represent a decisive step in the challenge of demonstrating benefits of improvement interventions through conventional healthcare methodologies (Shojania 2013). Second, this chapter provides a brief overview of affordances and illustrates the concept with a case study on the introduction of a new managerial technology. Third, it shows how the concept of affordances might contribute to understanding change in organizational practices and suggests implications for the study and reporting of QI interventions. Specifically for SQUIRE, additional attention shall be paid to the technical and interpretive properties of the intervention design and the recursive dynamics between QI intervention and context.

Issues in QI Implementation Research

Healthcare QI implementation research seeks to better understand the factors that affect the implementation and effectiveness of QI strategies. It is now accepted that an intervention that works in one setting does not necessarily work in another (Kaplan et al. 2010; Øvretveit 2011; Dixon-Woods et al. 2011). In this section, we examine two research initiatives that aim to systematically describe and analyse QI interventions, in order to improve control and replicability in QI research.

MUSIQ: An Attempt to Model the Moderating Impact of Context on QI Implementation and Outcomes

The Model for Understanding Success in Quality (MUSIQ) is a framework meant to facilitate research on contextual factors affecting QI implementation (Kaplan et al. 2010, 2012). It identifies twenty-five factors distributed in six overarching themes that reflect levels of analysis in the organization: external environment, organization, quality improvement capacity, clinical microsystem, quality improvement team and other miscellaneous issues (see Figure 2 in Kaplan et al. 2012, p. 17 for details). MUSIQ is distinct in focusing on clinical microsystem logics, which have been identified by an expert panel as critical (Kaplan et al. 2012).

The ultimate goal of the MUSIQ project is to gain predictive power in order to identify QI projects that are at risk of failure and provide guidance on actions that might improve results. This involves examining complex associations between context elements and QI success. Much of the empirical research on the role context plays in QI success has concentrated on individual relationships and has not examined more complex multifactorial associations or mediating relationships between aspects of context (Kaplan et al. 2013). Preliminary testing validated MUSIQ's reliability (Kaplan et al. 2013), but to our knowledge, no statistical work has been produced since the first exploratory study in 2013.

SQUIRE: An Attempt to Standardize QI Reporting

The Standards for Quality Improvement Reporting Excellence (SQUIRE) project is a publication guideline on how to report QI work in a systematic, reliable and consistent way. Most literature reviews of QI effectiveness underline that reports of improvement work vary widely in both content and quality (c.f., for example Grimshaw et al. 2004; Schouten et al. 2008), making it difficult to assess the determining conditions for success.

A first version of SQUIRE in 2008 sought to address this issue (Ogrinc et al. 2008). The more recent SQUIRE 2.0 emphasizes three key components of QI intervention reporting: the use of formal and informal theory in planning, implementing and evaluating improvement work; the context in which work is done; and the study of the intervention. The addition of an item termed 'rationale' is intended to clarify assumptions about the nature, context and expected outcomes of the intervention: authors are encouraged to explicitly report formal and informal theories about why they expected a particular intervention to work in a particular context (Ogrinc et al. 2015). Also, SQUIRE 2.0 recognizes context as a distinct item. Whilst it is rarely simple to isolate or describe context, understanding its impact on design, implementation, measurement and results is vital to identifying and reporting factors and mechanisms responsible for the success or failure of an intervention (Ogrinc et al. 2015; Goodman et al. 2016).

SQUIRE and MUSIQ aim to systematically describe and analyse QI interventions, and the context in which they are implemented, in order to enhance the replicability of QI studies. Both models emphasize the micro-level dynamics of the change process and consider recursive interactions between context and QI intervention.

However, based on organization and management literature insights, we suggest that further attention could be paid to the technical properties of QI interventions and the process of interaction with users. In the next section, we examine how the concept of affordances may shed light on these issues.

An Affordances Perspective on Practice Change

The notion of affordances originated in the field of psychology (Jones 2003), made its way into other social sciences, and has most recently appeared in organization and management studies. It is used to study the implementation of innovation and organizational change, and the role of managerial technologies and technical support in this process.

The concept of affordances sees the interaction between social processes and technical objects as key to understanding practice change: technical objects offer affordances that constrain and orient human action, whilst at the same time leaving some room for user discretion. In a nutshell, affordances are the *possibilities for action* offered by an object (Hutchby 2001). A technical object (artefact) possesses stable properties that endow it with specific possibilities for action; however, these possibilities must be perceived by a given actor to be effectively realized. The materiality of an object favours, shapes or invites, and at the same time constrains, a set of specific uses (Jarzabkowski and Kaplan 2014). For example, 'a chair offers a certain number of action possibilities: one can sit on it to rest, stand on it to gain height (…) However, a chair does not allow for certain actions. As it is, it cannot be used to fly, to dress oneself, or to be eaten as food' (Bérard 2014, p. 104).

Affordances first relate to the *artifact's materiality*. In the case of managerial technologies, *materiality* refers not only to material or technical properties, but also to interpretive properties that are embedded within the artefact and frame the array of possible interpretations by actors.

Affordances also relate to the *actor's interpretive capabilities*: the ability to understand and interpret the information at hand. 'The use depends not only on the material properties or on the intended design of the tool, but also on the context and the interpretations of actors who may use the technologies in creative, unpredictable ways' (Jarzabkowski and Kaplan 2014). This capability may vary according to the sociomaterial assemblage (Orlikowski and Scott 2008): embedded context characteristics such as the overall information system, actors' historical, cultural and professional references, and their views on the content and purpose of the implementation. Especially in managerial technologies, affordances may assume prior learning and skills in users (Hutchby 2001).

In this chapter, we retain two key propositions derived from the concept of affordances for application to QI initiatives as managerial technologies:

1. Change in clinical practice depends on the technology's technical and interpretive properties.
2. Change in clinical practice depends on the sociomaterial assemblage and the actors' interpretive capabilities.

Drawing on our previous research into the adoption of complex healthcare innovations, we will attempt to illustrate these two propositions and their implications for understanding models of QI implementation

Although there are plenty of case studies on QI implementation, they often lack a detailed description of contextual factors and the specificities of the QI intervention (Hoffmann et al. 2014). We, therefore, draw upon our research on the implementation of managerial technologies in healthcare settings, assuming that the inferences can be applied to other QI projects. We begin by providing an example of each proposition drawn from our research and then explore their methodological and theoretical implications for QI implementation studies.

Overall Context Description

In previous work, we studied the implementation of Operating Income Statements (OIS) in the medical divisions of a public hospital (Bérard 2014). At the time a disruptive change was underway in the French healthcare system, with the introduction of diagnostic-related group (DRG)-based hospital financing. In response, medical divisions were created in most hospitals, and OIS were introduced as the principal means of tracking their performance. OIS calculate the financial results of the medical division and are meant to help optimize performance, reveal pockets of productivity and act as an incentive to reduce expenses or increase revenues. According to hospital senior management, OIS should also, through benchmarking, promote emulation amongst medical divisions and create pressure to reduce costs. The OIS embody new financial rules for the new medical divisions, which have to reinvent themselves as if they were a collection of small private hospitals.

In practice, however, OIS are not being used as a decision-making tool to optimize financial performance. Rather, they are principally used by the finance department and physician heads of medical divisions as an *ex post* budget monitoring tool. They enable managers to register and explain expenses and revenues, *a posteriori*, three times a year, and provide a general overview of the distribution of resources within the hospital and amongst medical divisions. This use is not aligned with the dominant discourse of the actors within the hospital. Neither is it coherent with expectations expressed by institutions who publish guidance for OIS design and use in hospital settings (MeaH 2009).

How can the concept of OIS affordances help to understand this apparent discrepancy? We will see that it is due to both the technical and interpretive characteristics of the OIS, and to the sociomaterial assemblage in which OIS are embedded.

First Proposition: Change in Clinical Practice Depends on the Technology's Technical and Interpretive Properties

First, the OIS are complex in terms of presentation. OIS are comprehensive analytical tools, which present all medical division activities on a single page. The table routinely includes one hundred values grouped into three categories: direct expenses, indirect expenses and revenue. Each line of information is broken down into three columns: information for date n; information for date n−1; and difference n−1/n expressed as % change.

The information is derived from multiple interpretive operations and analyses in order to make it usable by participants. As is the case for any kind of income statement, the level of analysis chosen for presentation is just one of many possible options. The OIS table presents a specific value that reflects one level of interpretation whilst obscuring other possibilities. The figures are presented in terms of % change, suggesting a dynamic reading that favours comparison between one period and another. Indeed, users tend to focus on the '% change' column during meetings with the Finance Department, looking at changes over time and explaining variations in cost and revenue by focusing on changes in activity. In doing so, they neglect other types of analysis that might lead to different discussions.

Production of the OIS tables relies on multiple incremental calculations that stress the data processing system and obscure the process of data production. In the considered hospital case study, the information system was not sophisticated enough to meet the requirements of the OIS. Multiple operations—splitting or aggregating basic accounting units—are required to extract first-level information in a format compatible with the OIS. In addition, many of the values presented are the results of complex calculations that are carried out manually specifically for the production of OIS tables. This exposes the production of tables to a significant risk of calculation errors, as well as data manipulation and adjustment. Moreover, data are examined at an average delay of three months, reducing the possibilities for acting on their interpretation: examining June figures in late September makes it challenging to launch corrective actions that could rectify a trend for the year in question.

Second Proposal: Change in Clinical Practice Depends on the Sociomaterial Assemblage and the Actors' Interpretive Capabilities

During the study period, we observed that the configuration of the overall information system played an important role in the way OIS tables were interpreted. Indeed, the OIS is the first tool available to analyse medico-economic information at the medical division level. At the time the systems were created, no budget information was available on the monthly volume of resources allocated to and used by the medical divisions. Also, the production of data modelled on DRGs suffers important delays and is not easily understandable by doctors: for example, there is no direct way to translate DRG figures into more usual physician terms such as the volume of patients or types of disease.

The structure of decision-making and accountability amongst physician heads of the medical divisions also influences the way OIS are interpreted. When the OIS tables appeared, physician heads were made accountable for improvements in the medical division's net income, and incentives are based on these results. However, there is no formal negative consequence in case of failure. Physicians in public hospitals are often uncomfortable positioning themselves as managers when dealing with colleagues and administrators. The three-year executive position appointment contrasts with long-term relationships amongst medical peers. In addition, administrative managers are reluctant to give up power. For these reasons, the newly appointed heads of medical divisions can find it difficult to position themselves as decision-makers and use the OIS information to implement change.

Finally, use of the OIS is also strongly dependent on the physician's administrative and financial skills. Medical training in these areas is relatively weak, and there is a dearth of staff resources to help manage the information. Physicians receive the OIS tables during their quarterly meetings with the Finance Department, are given no time to study them beforehand, and are not encouraged to arrive at an in-depth understanding of the information.

Practical and Methodological Implications for QI Studies

Implications for QI: How Does This Study Relate to Prior Research?

Our study of the affordances of managerial technologies raises three important points. First, it suggests that managerial technologies cannot be considered as neutral devices, and that their interpretive properties need to be assessed. Managerial technologies structure the way quality issues are perceived. They offer representations of organizational reality, framing possibilities for and impediments to action: what they obscure is as important as what they highlight. Managerial techniques are not unequivocal or simple, but constitute a repertoire of possible interpretations. The local conditions for their production, the conventions on which they rely, and their presentation to end users are essential elements of the affordances of managerial technologies.

Second, the study emphasizes that managerial technologies have both expected and unexpected outcomes, and that both need to be considered when assessing implementation, remembering that 'success' can take different forms. However deviant from the managerial discourse, use of these technologies is still rational and produces effects on the organization, though perhaps not the intended effects.

Third, it underlines the embeddedness of contextual factors and managerial technologies. There is a dialectic interaction between technology and actors, within a sociomaterial assemblage comprising the actors' ability to interpret data, their hierarchical positioning, the way the technology operates within the information system and many other factors. Context matters, but it cannot be considered in isolation; there is an ongoing reconfiguration of context characteristics and managerial technologies. These observations inform and extend findings in prior research on QI implementation. First, the notion of a sociomaterial assemblage recalls the emphasis on contextual factors at microsystem level: clinical microsystems, QI teams and QI capacity (c.f., for example Pronovost et al. 2006, 2010; Kaplan et al. 2013). This calls for a reflection on the broad conceptual nature of context models, which currently emphasize multilevel structures, external environments and the organization and clinical practice at large (Kringos et al. 2015).

Second, the embeddedness of contextual factors and managerial technologies supports the idea of a dynamic and recursive evolution of QI interventions and context. This raises the issue of knowing when the intervention reaches maturity and stability, which is also addressed by MUSIQ and SQUIRE. 'The multiple relationships and pathways between exposure, outcome, and context variables in research on QI strategies are not yet sufficiently understood. Alternatively, context might be considered as an integral component of the subject area that evolves, changes and interacts with the intervention during the time period of QI project implementation. In this case, in-depth qualitative assessment is needed' (Kringos et al. 2015, p. 10). From an RCT perspective, Shojania (2013) also suggests that QI interventions should be studied only once they are stabilized, in order to ensure controllability and replicability.

Finally, the idea that managerial technologies have interpretive properties and can yield a variety of unintended outcomes suggests further insights for systematic descriptions of QI, as suggested below.

Methodological Implications: How to Study Affordances and Sociomaterial Assemblage

Echoing SQUIRE propositions (Ogrinc et al. 2015; Goodman et al. 2016), we suggest adding further elements to the reporting of QI interventions in order to better grasp the nature of the QI intervention and its impact on the implementation process (Table 3.1).

Table 3.1 Additions to reporting of QI interventions

SQUIRE categories	SQUIRE categories (details)	Proposed additions derived from affordances
Introduction	Specify rationale: The informal or formal frameworks, models, concepts and/or theories used to explain the problem and develop the intervention. 'Why did you think this would work'?	Specify the rationales and intended outcomes *according to each stakeholder* (QI team; clinician team; hospital management). 'What did you expect from this intervention'? Study each rationale *at both the beginning and end* of the study Compare the *intended rationale* with the QI's affordances and its *embedded rationale* (c.f. method section). 'How is the technology supposed to work'?
Method/context	Context is added as a vital contributor in identifying the mechanisms responsible for the success or failure of the intervention	Context should also be considered as a dependent variable. In particular, in longitudinal studies, the effects of the intervention on the context should be considered Include *governance aspects* (structure of decision-making and accountability for instance) in context
Method/intervention	Describe the intervention in sufficient detail that others could reproduce it, and specifics of the team involved in the work	Describe both the *technical attributes* and the *interpretive characteristics* of the technology E.g. 'What are the limitations and possible biases of the technology'? 'What proxies are used and how complex is their calculation and presentation'? 'Who performs the measurement'? 'When and how are results transmitted'? 'How are they presented and how easy are they to understand'?

(continued)

Table 3.1 (continued)

SQUIRE categories	SQUIRE categories (details)	Proposed additions derived from affordances
Results	Describe the initial steps of the intervention and their evolution over time, as well as observed associations between outcomes, interventions and relevant contextual elements. Unintended consequences will also be reported	Include an additional description of the way QI initiative is used: What use is made of the information at hand and in what kind of activities is QI involved: control (goal fixation, measurement, performance evaluation, decision-making); care provision (care activities and decision-making)

In conclusion, this chapter illustrates how the organization and management literature can contribute valuable insights to the study of QI implementation in healthcare settings. Both strands of literature emphasize the importance of context and microsystem dynamics in understanding and potentially replicating QI implementation success. Moreover, an affordance perspective suggests a need to pay more attention to the QI design itself and its enabling and constraining potential for practice change.

References

Bate, P., Robert, G., Fulop, N., Øvretveit, J., & Dixon-Woods, M. (2014). *Perspectives on context*. London: The Health Foundation.

Bérard, É. (2014). Management control artefacts: An enabling or constraining tool for action? Questioning the definition and uses of the concept of affordances from a management control perspective. In F.-X. de Vaujany, N. Mitev, P. Laniray, & E. Vaast (Eds.), *Materiality and time* (pp. 99–123). New York: Springer.

Dixon-Woods, M., Bosk, C. L., Aveling, E. L., Goeschel, C. A., & Pronovost, P. J. (2011). Explaining Michigan: Developing an ex post theory of a quality improvement program. *Milbank Quarterly, 89*(2), 167–205.

Goodman, D., Ogrinc, G., Davies, L., Baker, G. R., Barnsteiner, J., Foster, T. C., Gali, K., Hilden, J., Horwitz, L., & Kaplan, H. C. (2016). Explanation and elaboration of the SQUIRE (Standards for Quality Improvement Reporting Excellence) Guidelines, V. 2.0: Examples of SQUIRE elements in the healthcare improvement literature. *BMJ Quality & Safety, 25*(12), e7–e7.

Grimshaw, J. M., Thomas, R. E., MacLennan, G., Fraser, C., Ramsay, C. R., Vale, L., Whitty, P., Eccles, M. P., Matowe, L., Shirran, L., Wensing, M., Dijkstra, R., & Donaldson, C. (2004). Effectiveness and efficiency of guideline dissemination and implementation strategies. *Health Technology Assessment, 8*(6), 1–72.

Hoffmann, T. C., Glasziou, P. P., Boutron, I., Milne, R., Perera, R., Moher, D., Altman, D. G., Barbour, V., Macdonald, H., & Johnston, M. (2014). Better reporting of interventions: Template for intervention description and replication (TIDieR) checklist and guide. *British Medical Journal, 348*, g1687.

Hutchby, I. (2001). Technologies, texts and affordances. *Sociology, 35*(2), 441–456.

Jarzabkowski, P., & Kaplan, S. (2014). Strategy tools-in-use: A framework for understanding 'technologies of rationality' in practice. *Strategic Management Journal, 36*(4), 537–558.

Jones, K. S. (2003). What is an affordance? *Ecological Psychology, Special Issue on Affordances, 15*(2), 107–114.

Kaplan, H. C., Brady, P. W., Dritz, M. C., Hooper, D. K., Linam, W., Froehle, C. M., & Margolis, P. (2010). The influence of context on quality improvement success in health care: A systematic review of the literature. *Milbank Quarterly, 88*(4), 500–559.

Kaplan, H. C., Froehle, C. M., Cassedy, A., Provost, L. P., & Margolis, P. A. (2013). An exploratory analysis of the model for understanding success in quality. *Health Care Management Review, 38*(4), 325–338.

Kaplan, H. C., Provost, L. P., Froehle, C. M., & Margolis, P. A. (2012). The Model for Understanding Success in Quality (MUSIQ): Building a theory of context in healthcare quality improvement. *BMJ Quality & Safety, 21*(1), 13–20.

Kringos, D. S., Sunol, R., Wagner, C., Mannion, R., Michel, P., Klazinga, N. S., & Groene, O. (2015). The influence of context on the effectiveness of hospital quality improvement strategies: A review of systematic reviews. *BMC Health Services Research, 15*(1), 277.

MeaH (ANAP). (2009). Nouvelle gouvernance et comptabilité analytique par pôles. CREA, CREO, TCCM, tableaux de bord, une aide méthodologique au dialogue de gestion.

Ogrinc, G., Davies, L., Goodman, D., Batalden, P., Davidoff, F., & Stevens, D. (2015). SQUIRE 2.0 (Standards for QUality Improvement Reporting Excellence): Revised publication guidelines from a detailed consensus process. *The Journal of Continuing Education in Nursing, 46*(11), 501–507.

Ogrinc, G., Mooney, S. E., Estrada, C., Foster, T., Goldmann, D., Hall, L. W., Huizinga, M. M., Liu, S. K., Mills, P., & Neily, J. (2008). The SQUIRE (Standards for QUality Improvement Reporting Excellence) guidelines for quality improvement reporting: Explanation and elaboration. *Quality and Safety in Health Care, 17*(Suppl 1), i13–i32.

Orlikowski, W. J., & Scott, S. (2008). Sociomateriality: Challenging the separation of technology, work and organization. *Annals of the Academy of Management, 2*(1), 433–474.

Øvretveit, J. (2011). Understanding the conditions for improvement: Research to discover which context influences affect improvement success. *BMJ Quality & Safety, 20*(Suppl 1), i18–i23.

Pronovost, P., Goeschel, C. A., Colantuoni, E., Watson, S., Lubomski, L. H., Berenholtz, S. M., Thompson, D. A., Sinopoli, D. J., Cosgrove, S., & Sexton, J. B. (2010). Sustaining reductions in catheter related bloodstream infections in Michigan intensive care units: Observational study. *British Medical Journal, 340*, c309.

Pronovost, P., Needham, D., Berenholtz, S., Sinopoli, D., Chu, H., Cosgrove, S., Sexton, B., Hyzy, R., Welsh, R., & Roth, G. (2006). An intervention to decrease catheter-related bloodstream infections in the ICU. *New England Journal of Medicine, 355*(26), 2725–2732.

Schouten, L. M. T., Hulscher, M. E. J. L., van Everdingen, J. J. E., Huijsman, R., & Grol, R. P. T. M. (2008). Evidence for the impact of quality improvement collaboratives: Systematic review. *British Medical Journal, 336*(7659), 1491–1494.

Shojania, K. G. (2013). *Conventional evaluations of improvement interventions: More trials or just more tribulations?* London: BMJ Publishing Group Ltd.

Stevens, D. P., & Shojania, K. G. (2011). Tell me about the context, and more BMJ Quality & Safety, 20(7), 557–559.

Tomoaia-Cotisel, A., Scammon, D. L., Waitzman, N. J., Cronholm, P. F. Halladay, J. R., Driscoll, D. L., Solberg, L. I., Hsu, C., Tai-Seale, M., & Hiratsuka, V. (2013). Context matters: The experience of 14 research team in systematically reporting contextual factors important for practice change The Annals of Family Medicine, 11(Suppl 1), S115–S123.

4

Emerging Hybridity: A Comparative Analysis of Regulatory Arrangements in the Four Countries of the UK

Joy Furnival, Ruth Boaden and Kieran Walshe

Introduction

This chapter outlines a study that aims to understand and analyse the different regulatory models in the UK, by identifying regulatory model developments and challenges. This chapter begins by detailing an analysis framework built on regulatory theoretical concepts. Next, the scope and methods for the study are detailed, followed by a description of the regulatory architecture across the UK. This outlines an emerging trend towards hybrid models of regulation. The tensions that emerge from

This chapter is adapted from a previously published article available open access under a CC BY 4.0 license at https://doi.org/10.1108/JHOM-06-2016-0109.

J. Furnival (✉)
NHS Improvement, London, UK
e-mail: Joy.furnival1@nhs.net

R. Boaden · K. Walshe
Alliance Manchester Business School, University of Manchester, Booth Street East, Manchester M13 9SS, UK

© The Author(s) 2018
A.M. McDermott et al. (eds.), *Managing Improvement in Healthcare*,
Organizational Behaviour in Health Care,
https://doi.org/10.1007/978-3-319-62235-4_4

59

this development are described in the findings and discussion. This chapter ends by indicating the contribution to research and practice.

Healthcare Regulation

Selznick (1985) defines regulation as 'sustained and focused control exercised by a public agency over activities which are valued by a community' (p. 363). Regulation occurs for several reasons including protection from market failures, critical goods shortages and moral hazards (Feintuck 2012). In healthcare, regulation is used to address demands for improved performance.

Regulation is described as having three aims, accountability, assurance and improvement (Walshe 2003b) which can be delivered through three regulatory models. These are deterrence, compliance and responsive (Ayres and Braithwaite 1992; Reiss 1984). Bardach and Kagan (1982) indicate that deterrence models assume that organisations are amoral and will flout rules deliberately if they are not enforced, whereas compliance models assume that organisations try to 'do the right thing', but events occur and things go wrong, and organisations will need support to resolve issues. Responsive regulatory models use a combination of deterrence and compliance models contingent on the local circumstances (Ayres and Braithwaite 1992; Braithwaite 2011). Responsive regulatory agencies are described as hybrids that use both deterrence and compliance models concurrently to ensure improvement (McDermott et al. 2015).

Regulatory agencies use three processes, direction, detection and enforcement, which can be used with different levels of emphasis (Walshe 2003b). Direction incorporates the setting and influencing of standards, guidance and policy. Detection includes inspection, measuring and monitoring of performance. Finally, enforcement includes a range of methods used to encourage and force behavioural change, such as sanctions, education and support (Walshe and Shortell 2004; Hutter 1989). Regulation requires standards to be maintained and provides valuable feedback for improvement (Gunningham 2012). Nevertheless, regulation is critiqued for many reasons, including high costs (Ng 2013), inflexibility (Brennan 1998), tunnel vision (Mannion et al. 2005), ineffectiveness (Flodgren et al. 2011), inhibiting innovation (Stewart 1981), capture (Boyd and Walshe 2007) and ritualistic compliance (Braithwaite et al. 2007).

Given these criticisms, new regulatory models are increasingly proposed using professionalism and improvement support (Ham 2014) concurrently with other regulatory methods such as inspection. These 'hybrids' Fischer and Ferlie can be viewed as a variant of responsive regulation to ensure improved performance. However, (Fischer and Ferlie 2013) argue that regulatory models consist of various values and norms which cannot be readily combined. One of the few studies that analyse the influence and impact of regulatory hybridity suggest that without a receptive context, collaborative stakeholder relationships, adequate resources and time, regulatory responsibility for improvement approaches may need to be separated (McDermott et al. 2015). This chapter contributes by comparing regulatory models across the UK to understand the tensions within hybrid regulatory models.

Methodology

In the UK, there are six organisational regulatory agencies. These are the Regulatory and Quality Improvement Authority (RQIA) in Northern Ireland, Healthcare Improvement Scotland (HIS), the Healthcare Inspectorate Wales (HIW), and in England, Monitor, the Care Quality Commission (CQC) and the Trust Development Authority (TDA). These agencies all review acute hospital-based care, which accounts for the majority of UK healthcare expenditure, enabling comparison. The six organisations were approached to take part in the study and all agreed. Ethical permission to proceed with the study was also received.

Policy documents were identified from regulatory agencies that included information connected to regulatory aims, strategy and results, and were analysed alongside anonymous transcripts from forty-eight semi-structured interviews of a cross section of staff from each agency. The interviews were conducted between October 2014 and April 2015, and participation was confidential and voluntary. Five pilot interviews were conducted, and finalised questions included 'what is the aim of this regulatory agency?', and 'what enforcement interventions are used?'. Thematic analysis (Boyatzis 1998) was used to analyse and compare the regulatory models using an a priori framework identified from the literature.

Findings

The findings are presented in three sections covering the regulatory architecture in the UK, the regulatory models and the challenges identified.

Northern Ireland

Following devolution, the RQIA was established in 2005. As the main regulatory agency for health and social care services in Northern Ireland, it employs 152 staff and has a budget of £7.6 M (2013/2014). It aims to regulate, scrutinise and drive improvement in services for a population of 1.8 M.

Scotland

HIS was established in 2011 following a merger with several predecessor organisations. It aims to advance healthcare improvement in Scotland and to ensure the delivery of safe, effective and person-centred care for a population of 5.3 M. It does not review social care. It has a budget of £20 M (2014/2015) and employs 329 staff.

Wales

HIW was established following devolution in 2004, as a unit within the Welsh Assembly Government. It is responsible for the inspection of health services including General Practitioner practices, pharmacies and dental practices, but like HIS, not social care. It has a budget of £3 M and fifty-nine staff to oversee health services for a population of 3 M.

England

Three regulatory agencies oversee healthcare services within England. This contributes to a more fragmented and yet overlapping landscape. All three agencies have responsibilities to review acute care, mental health, community and ambulance services but with slightly different scope.

The CQC was formed in 2009 from a merger of predecessor organisations in England, including a former social care inspectorate. The CQC's purpose is to ensure high-quality care is provided and to encourage improvement (Care Quality Commission 2013). It uses inspection to cover a wide range of services across different health services including dentistry, primary, mental, community, acute and social care, covering over 56,000 individual delivery locations (Care Quality Commission 2015). It has 2581 employees and a budget of £240 M (2014/2015) covering health and social care services for an English population of 53 M.

Since the early 2000s, the English National Health Service (NHS) has been encouraging the development of Foundation Trusts (Walshe 2003a). Foundation Trusts are accountable to local people and can decide locally how to meet their obligations, rather than this being decided by the Department of Health. Monitor was established in 2004 to oversee Foundation Trusts in England, and it is a non-departmental public body of the Department of Health. Following the Health and Social Care Act in 2012 (HSCA), its role includes price setting, preventing anti-competitive behaviour and regulating finances, quality and performance for approximately 149 Foundation Trusts. Concurrently, it is required to promote care service integration and protect services for patients in the event of organisational unsustainability. It employees 532 staff and has a budget of £72 M (2014/2015).

The TDA is a special health authority of the Department of Health, set up following the HSCA. It fulfils a similar role as Monitor for approximately ninety non-Foundation Trusts and is responsible for developing them into Foundation Trusts. It does not hold formal regulatory powers. It employs 315 staff with a budget of £65 M

(2014/2015). There is some explicit overlap of the regulatory responsibilities across the three regulatory agencies in England, particularly for oversight of care quality and governance.

Regulatory Agency Comparison

The six regulatory agencies all oversee acute, community, mental health and ambulance care within their respective country. However, substantial differences also exist; for example, HIS, HIW, Monitor and the TDA do not oversee social care whereas the CQC and RQIA do. HIS, Monitor and TDA provide improvement support, whereas the others do not. Some also have niche responsibilities, such as HIW for pharmacies and RQIA for commissioning. Therefore, each agency has different volumes, types and scope of organisations to oversee, covering different populations. For example, the TDA only oversees approximately ninety NHS organisations, whereas the CQC reviews services in over 56,000 locations (Care Quality Commission 2015). This makes it difficult to find a common denominator for comparison.

Nevertheless, some comparison can be made. First, RQIA has over double the budget of HIW for a smaller population, reflecting RQIA's wider scope in the oversight of social care. Second, HIWs budget seems small when compared with HIS even when population and scope differences are accounted for. In England, the CQC reviews over 56,000 locations with a budget of £240 M, whilst Monitor and TDA have significantly fewer organisations to review, yet their combined expenditure in 2014/2015 was over half that of the CQC. Further, the TDA spends approximately £237 k more per organisation than Monitor (£485 k/organisation versus £722 k/organisation), perhaps reflecting greater financial support provided to non-Foundation Trusts.

Table 4.1 analyses the documents and interviews to compare regulatory goals and models. Categorising the regulatory model for each agency was not simple as agencies may demonstrate aspects of several regulatory models. The term 'hybrid' is used to illustrate an emergent responsive regulatory approach whereby regulatory agencies are primarily using enforcement methods that comprise improvement support

Table 4.1 Agency goals and models

Agency	Documentary data	Interview data	Regulatory model
HIS	'We are the national healthcare improvement organisation for Scotland, established to advance improvement in healthcare' (Healthcare Improvement Scotland 2014a)	'...a blend of approaches, so we have the scrutiny, assurance, we have the clinical expertise ... independent fair and objective assessment ... [and] ... support improvement efforts' (Interview participant G, HIS) '[we]... help providers in Scotland to improve their improvement capability' (Interview participant A, HIS)	Hybrid
HIW	'Our purpose is to provide independent and objective assurance on the quality, safety and effectiveness of healthcare services, making recommendations to healthcare organisations to promote improvements' (Healthcare Inspectorate Wales 2014a)	'We go out and inspect and we find ... an organisation is meeting the standards ... then we wouldn't seek improvement ... beyond that' (Interview participant B, HIW) 'We are not an improvement agency, but we should be operating in a way which supports improvement' (Interview participant D, HIW)	Compliance
RQIA	'The most important priority for RQIA is to make sure that our inspection systems and processes convey clearly to the public how well a service is performing in respect of the ... minimum standards' (Regulation and Quality Improvement Authority 2015b)	'We provide assurance ... about the quality of services' (Interview participant D, RQIA) 'Our primary role is to question them, to challenge them early, and then they can then start making ... improvements' (Interview participant A, RQIA)	Compliance
CQC	'We make sure health and social care services provide people with safe, effective, compassionate, high-quality care and we encourage care services to improve' (Care Quality Commission 2013)	'We monitor, we inspect and we regulate and make sure that these services meet the fundamental standards' (Interview participant CQC D) 'It's very clear in the CQC that we're not improvement facilitators, we're regulators' (Interview participant C, CQC)	Compliance

(continued)

Table 4.1   (continued)

Agency	Documentary data	Interview data	Regulatory model
Monitor	'[We set] a required standard that all NHS providers must meet ... [We] control the risk that foundation trusts, once authorised, fall back below the required standard. If they do, we take remedial action ... We will focus in particular on the capabilities that drive long-term performance' (Monitor 2014)	'Where trusts fail to deliver certain minimum standards ... [we] work with those trusts to ensure that they improve their position and restore themselves to ... that minimum standard' (Interview participant A, Monitor) '[Our] mandate is basically to improve the capability of FTs' (Interview participant G, Monitor)	Hybrid
TDA	'The TDA oversees NHS trusts and holds them to account ... while providing them with support to improve' (Trust Development Authority 2014)	'[Trusts] know that they are being held to account for their performance but they also know that they will get support and help and development rather than just being criticised' (Interview participant G, TDA) '[Our role is] supporting oversight of our Trusts, ... [and] that have asked for some support because they feel that they need to make some improvements' (Interview participant E, TDA)	Hybrid

through direct action, and that this is tailored contingent on organisational circumstances and performance. Three agencies met these criteria, and were categorised as 'hybrid' regulatory agencies. The remaining agencies described methods that remained unchanged, regardless of organisational circumstances and because the enforcement methods used did not include the provision of improvement support.

This demonstrates that the agencies have similar goals to improve and assure care. Analysis of the documents and interviews indicates that there are differing models and methods used, shown in Table 4.2.

Table 4.2 shows how the agencies use some form of assessment to review care provision including self-assessment (all), formal inspection (RQIA, CQC, HIS, HIW), and thematic reviews (RQIA, CQC, HIS,

Table 4.2 Agency methods

Tactic	Method	HIS	HIW	RQIA	CQC	Monitor	TDA
Detect	Unannounced inspections	X			X		
	Planned inspections/ reviews	X	X	X	X	X	
	Thematic, specialist and peer review	X	X	X	X		
	Performance monitoring	X	X	X	X	X	X
	Self-assessment/ declaration	X	X	X	X	X	X
Direct	Develops standards	X			X	X	
Enforce	Provides improvement support through direct action	X				X	X
	Highlights best practice	X	X	X	X	X	X
	Formal powers	Independent care only	Independent care only	X	X	X	

HIW). The CQC, HIS and Monitor develop standards, whilst others provide improvement support (HIS, Monitor, TDA). Three agencies do not have formal powers for NHS organisations (HIS, RQIA, TDA). This shows the dominance of compliance activities within regulatory agencies, and that improvement activity is often limited to the promotion of best practices. Half the agencies consider their role to be providing public assurance and use similar methods regardless of performance or risk (CQC, HIW, RQIA), meeting the description of compliance models. The remaining three agencies (Monitor, TDA, HIS) all described enforcement methods including education and improvement support through programmes such as the Scottish Patient Safety Programme (Healthcare Improvement Scotland 2014b) as well as other enforcement action that was contingent on specific circumstances, indicating the use of hybrid models of regulation.

Tensions Within Hybrid Models

The analysis highlights tensions caused by the combination of assurance, accountability and improvement goals.

> …it's quite clear that we're there to scrutinise and to regulate, but we're also there to try to help improvement … it isn't always easy to fit the two together (Interview participant H, CQC)

> [NHS] Boards are saying, actually, don't confuse us. You can't come in with an inspection hat on and then an improvement one (Interview participant C, HIS)

This chapter identifies three themes from these tensions: regulatory roles, resources and relationships.

Regulatory Roles

Interview participants and agency documents describe a tension between their roles to assure and improve care.

> Quality care cannot be achieved by inspection and regulation alone. The main responsibility for delivering quality care lies with [those that provide], arrange and fund local services (Care Quality Commission 2013)

> The Berwick report (2013) highlights the vital role that 'intelligent inspection' plays. However, this cannot stand alone and must be combined within a system of improvement (Healthcare Improvement Scotland 2014a)

> We're very clear what our role is when we go in, and our role is not to run the trust or run a piece of work (Interview participant A, TDA)

Some agencies were concerned that delivering improvement activity compromises their 'role' to conduct objective detection. Interview participants also raised concerns regarding accountability should the improvement support not lead to the expected outcomes.

there is a danger of conflict, that we mark our own homework ... a hospital [could] say, but you've been working with us on this so the failure is also partly yours (Interview participant A, Monitor)

When trusts aren't performing, there is a lot of pressure in the system, to say ... to almost indicate that it's wilful. It's almost as if they're failing for reasons which they should be able to stop (Interview participant C, TDA)

We don't make standards because it would be an uncomfortable place to be, to be the regulator and review against your own standards (Interview participant E, RQIA).

Resources

The choice of regulatory model has ramifications for planning and execution, as it affects the type of resources (e.g. information technology versus clinical skills) that are needed, and influences the resources available for other tasks. For example, compliance models need more inspectors, whereas hybrid models need more improvement facilitators. This makes the choice of regulatory model more path dependant and slows to change. Analysis reveals that that few employees have improvement skills or experience within regulating agencies. Shortages need addressing through development, recruitment and investment.

We had no resources to take it forward (Interview participant B, HIS)

We've got quite a big, sort of, issue about needing to invest in our staff ... you can't just outsource ... we just don't have the time and need some supplemental space to be able to really engage with [improvement] (Interview participant A, CQC)

There [is] a challenge to find people of those skills (Interview participant B, HIW)

It is clear from the documents and interviews that some participants resisted these developments. This is partly due to the lengthy period and high costs of developing skills. It also links to disagreements regarding the regulatory aims and due to concerns regarding local accountability.

[I wonder] how knowledgeable the inspectors are around improvement methodology because you can't judge it unless you know what you're looking for ... I think the inspectors lack the improvement methodology understanding ... we don't have the special advisors either (Interview participant C, CQC)

We haven't got anything like the number of people working within Monitor that have the [improvement] experience they'd need ... some people would say, this isn't a job for a regulator (Interview participant F, Monitor)

RQIA has limited capacity [...] to encourage service providers to continuously improve (Regulation and Quality Improvement Authority 2015a).

Regulatory agencies report pressures linked to resources and describe a trade-off required between detection and enforcement activities and the resources available.

...we would have to think carefully about whether our time's better spent doing [improvement work] or another inspection somewhere else (Interview participant B, HIW)

...with regulation, you have to prioritise. If we were regulating everybody it wouldn't have any impact and [we] wouldn't have enough resources (Interview participant B, Monitor)

Relationships

The final theme is relationships. Interview participants commented on their need to maintain objectivity and prevent regulatory capture to assure the public that their assessments of care quality were fair, trustworthy and accurate. However, interview participants acknowledged the risks of negative reporting, noting that detection and enforcement together with tough media and political scrutiny can develop a destabilising effect on organisations and associated relationships.

...if you establish good ongoing relationships outside the inspection regime, then it's less about you coming in and more about the team that the hospital knows (Interview participant C, CQC)

... You're still having that professional distance as a regulator but you get to know the chief exec ... and they get to know you (Interview participant F, HIW)

the approach of some providers might be ... they're a regulator so I don't want to go near them, whereas some of our best relationships with trusts are ... coming to us very early for advice (Interview participant B, Monitor)

However, analysis indicated that agencies believe that enforcement action, both punitive and supportive, must be transparent to prevent against regulatory capture to maintain public trust in 'independent and objective' regulatory agencies.

HIW will report clearly, openly and publicly on the work that we undertake in order that citizens are able to access independent and objective information on the quality, safety and effectiveness of healthcare in Wales (Healthcare Inspectorate Wales 2014a, b).

By publicly reporting our findings, we provide assurance to the public that standards are being met, or that action is being taken where improvements are needed (Healthcare Improvement Scotland 2013)

These two contrasting perspectives, of confidentiality and openness, are more difficult to reconcile.

There is an inherent tension with that confidential, closed-doors enquiry support with the requirements for us as a body about public accountability and transparency (Interview participant G, HIS)

Finally, external stakeholders, such as the media, may use information differently, hindering relationship development, mutual trust and care improvement in some circumstances. Those providing care may be concerned that information disclosure may deter honest discussion of problems due to these stakeholders (Berwick et al. 2003).

Discussion and Implications

The study described in this chapter aims to explore the regulatory architecture and models across the devolved countries of the UK. It describes how regulatory agencies have differing scope and methods to deliver their goals. The analysis presented within this chapter illuminates how the ability of regulatory agencies to balance their requirements to assure and improve care relates to effective regulatory oversight. In response, hybrid regulatory models are emerging within three of the agencies (TDA, Monitor and HIS).

Hybridity is a concept that is used widely to describe organisational responses to governance changes supporting the use of different organisational models to satisfy multiple demands (Skelcher and Smith 2015; Miller et al. 2008). However, it can lead to identity disruption and unstable organisations with contradictory organisational goals that cannot be easily combined (Denis et al. 2015; Skelcher and Smith 2015; Smith 2014). To manage the tensions within hybrid regulatory models identified within this study, regulatory agencies may find it helpful to clarify the relationship between accountability, assurance and improvement by articulating their improvement model and regulatory role (Davidoff et al. 2015).

This clarification may also ensure the relevant regulatory and improvement skills are recruited, reducing tensions developing through a lack of appropriately skilled staff. The model could also reduce strained regulatory relationships with organisations and a potential organisational dependency on the improvement support by clarifying organisational and regulatory roles and intentions.

Conclusion

Hybrid regulatory models are emerging in the UK. These supplement deterrence and compliance enforcement methods by using direct improvement support with healthcare organisations. However, the execution of these emerging hybrid models is complex and emergent. Three areas of tensions are identified when developing hybrid models:

regulatory roles, resources and relationships. Effective healthcare regulation requires recognition of the inherent tensions between the regulatory aims of accountability, assurance and improvement, and clarity of their intended connections.

References

Ayres, I., & Braithwaite, J. (1992). *Responsive regulation: Transcending the deregulation debate.* New York: Oxford University Press.

Bardach, E., & Kagan, R. (1982). *Going by the book: The problem with regulatory unreasonableness—A twentieth century fund report.* Philadelphia, PA: Temple University Press.

Berwick, D. (2013). *A promise to learn—A commitment to act: Improving the safety of patients in England.* London: National Advisory Group on the Safety of Patients in England.

Berwick, D. M., James, B., & Coye, M. J. (2003). Connections between quality measurement and improvement. *Medical Care, 41*(1), 130–138.

Boyatzis, R. (1998). *Transforming qualitative Information: Thematic analysis and code development.* Thousand Oaks, CA: Sage.

Boyd, A., & Walshe, K. (2007). *Designing regulation: A review. A review for the Healthcare Commission of the systems for regulation and their impact on performance in seven sectors, and a synthesis of the lessons and implications for regulatory design in healthcare.* Manchester: Manchester Business School.

Braithwaite, J. (2011). The essence of responsive regulation. *U.B.C. Law Review, 44*(3), 475–520.

Braithwaite, J., Makkai, T., & Braithwaite, V. (2007). *Regulating aged care: Ritualism and the new pyramid.* Cheltenham: Edward Elgar Publishing Ltd.

Brennan, T. A. (1998). The role of regulation in quality improvement. *The Milbank Quarterly, 76*(4), 709–731.

Care Quality Commission. (2013). *Raising standards, putting people first: Our strategy for 2013 to 2016.* London: Care Quality Commission.

Care Quality Commission. (2015). *Services we monitor, inspect and regulate.* London: Care Quality Commission. Retrieved from http://www.cqc.org.uk/. Accessed 16 June 2015.

Davidoff, F., Dixon-Woods, M., Leviton, L., & Michie, S. (2015). Demystifying theory and its use in improvement. *BMJ Quality & Safety, 0*, 1–11. doi: 10.1136/bmjqs-2014-003627.

Denis, J. L., Ferlie, E., & Van Gestel, N. (2015). Understanding hybridity in public organizations. *Public Administration, 93*(2), 273–289.

Feintuck, M. (2012). Regulatory rationales beyond the economic: In search of the public interest. In R. Baldwin, M. Cave, & M. Lodge (Eds.), *The Oxford handbook of regulation*. Oxford: Oxford University Press.

Fischer, M. D., & Ferlie, E. (2013). Resisting hybridisation between modes of clinical risk management: Contradiction, contest, and the production of intractable conflict. *Accounting, Organizations and Society, 38*(1), 30–49.

Flodgren, G., Pomey, M. P., Taber, S. A., & Eccles, M. P. (2011). Effectiveness of external inspection of compliance with standards in improving healthcare organisation behaviour, healthcare professional behaviour or patient outcomes. *Cochrane Database of Systematic Reviews, 11*.

Gunningham, N. (2012). Enforcement and compliance strategies. In R. Baldwin, M. Cave, & M. Lodge (Eds.), *The Oxford handbook of regulation*. Oxford: Oxford University Press.

Ham, C. (2014). *Reforming the NHS from within: Beyond hierarchy, inspection and markets*. London: The King's Fund.

Healthcare Improvement Scotland. (2013). *Annual Report 2012–2013*. Edinburgh: Healthcare Improvement Scotland.

Healthcare Improvement Scotland. (2014a). *Driving Improvement in Healthcare. Our Strategy 2014–2020*. Edinburgh: Healthcare Improvement Scotland.

Healthcare Improvement Scotland. (2014b). *Highlights of the Scottish patient safety programme national conference—Driving improvements in patient safety*. Edinburgh: Healthcare Improvement Scotland.

Healthcare Inspectorate Wales. (2014a). *Annual Report 2013–2014*. Cardiff: Healthcare Inspectorate Wales.

Healthcare Inspectorate Wales. (2014b). *Healthcare inspectorate wales operational plan 2014–2015*. Merthyr Tydfil: Healthcare Inspectorate Wales.

Hutter, B. M. (1989). Variations in regulatory enforcement styles. *Law & Policy, 11*(2), 153–174.

Mannion, R., Davies, H., & Marshall, M. (2005). Impact of star performance ratings in English acute hospital trusts. *Journal of Health Services Research & Policy, 10*(1), 18–24.

McDermott, A. M., Hamel, L. M., Steel, D. R., Flood, P. C., & McKee, L. (2015). Hybrid healthcare governance for improvement? Combining top-down and bottom-up approaches to public sector regulation. *Public Administration, 93*(2), 324–344.

Miller, P., Kurunmäki, L., & O'Leary, T. (2008). Accounting, hybrids and the management of risk. *Accounting, Organizations and Society, 33*(7–8), 942–967.

Monitor. (2014). *Monitor's strategy 2014–2017: Helping to redesign healthcare provision in England.* London: Monitor.

Ng, G. K. B. (2013). Factors affecting implementation of accreditation programmes and the impact of the accreditation process on quality improvement in hospitals: A SWOT analysis. *Hong Kong Medical Journal, 2013*(19), 434–436.

Regulation and Quality Improvement Authority. (2015a). *Board papers January 2015.* Belfast: Regulation and Quality Improvement Authority.

Regulation and Quality Improvement Authority. (2015b). *Corporate strategy 2015–2018.* Belfast: Regulation and Quality Improvement Authority.

Reiss, A. J. (1984). Selecting strategies of social control over organisational life. In K. Hawkins & J. M. Thomas (Eds.), *Enforcing regulation.* Boston, MA: Kluewer-Nijhoff.

Selznick, P. (1985). Focusing organizational research on regulation. In R. G. Noll (Ed.), *Regulatory policy and the social sciences.* Berkeley, CA: University of California Press.

Skelcher, C., & Smith, S. R. (2015). Theorizing hybridity: Institutional logics, complex organizations, and actor identities: The case of non-profits. *Public Administration, 93*(2), 433–448.

Smith, S. R. (2014). Hybridity and nonprofit organizations: The research agenda. *American Behavioral Scientist, 58*(11), 1494–1508.

Stewart, R. B. (1981). Regulation, innovation, and administrative law: A conceptual framework. *California Law Review, 69*(5), 1256–1377.

Trust Development Authority. (2014). *Delivering for patients: The 2014/2015 accountability framework for NHS trust boards.* London: Trust Development Authority.

Walshe, K. (2003a). Foundation hospitals: A new direction for NHS reform? *Journal of the Royal Society of Medicine, 96*(3), 106–110.

Walshe, K. (2003b). *Regulating healthcare: A prescription for improvement?* Maidenhead: Open University Presss.

Walshe, K., & Shortell, S. M. (2004). Social regulation of healthcare organizations in the United States: Developing a framework for evaluation. *Health Services Management Research, 17*(2), 79–99.

5

Contextual Factors Affecting the Implementation of Team-Based Primary Care: A Scoping Review

Dori A. Cross

Introduction

Comprehensive primary care is the cornerstone of a low cost, accessible and high-quality health system, and robust primary care infrastructure is a key to tackling unsustainable growth in health spending and significant gaps in patient care quality and outcomes (Donaldson et al. 1996; Starfield et al. 2005). Strengthened inter-professional teamwork amongst primary care physicians and practice staff—including nurses, medical assistants (MAs) and others—has emerged as a promising strategy to promote more effective care delivery, particularly as concurrent delivery reforms such as patient-centred medical home (PCMH) and pay-for-performance initiatives seek to expand the scope of primary care services. Team-based primary care (TBPC) can alleviate mounting time pressures on primary care physicians through improved delegation and

D.A. Cross (✉)
University of Michigan School of Public Health,
Ann Arbor, MI, USA
e-mail: dacross@umich.edu

© The Author(s) 2018   **77**
A.M. McDermott et al. (eds.), *Managing Improvement in Healthcare*,
Organizational Behaviour in Health Care,
https://doi.org/10.1007/978-3-319-62235-4_5

empowerment of other staff members to work to the fullest extent of their training (Friedman et al. 2014; Mitchell et al. 2012; Saba et al. 2012; Shipman and Sinsky 2013). Furthermore, there is a reasonable body of empirical evidence supporting the link between the adoption of TBPC and improved efficiency (Page 2006; Thomas 2014), quality and comprehensiveness of services (Cutrona et al. 2010; McAllister et al. 2013; Mohr et al. 2013; Roblin et al. 2011), and physician, staff and patient satisfaction (Altschuler et al. 2012; Helfrich et al. 2014; Willard-Grace et al. 2014). Indeed, for many practices, adopting a team-based approach to care may be the critical element needed to realize the intended benefits of broader care reform efforts such as PCMH transformation (McAllister et al. 2013).

However, these relationships are not iron-clad. Estimates of the positive effects of TBPC on outcomes have been found to be unreliable, with calls for research to improve understanding of the mechanisms and facilitating factors which help practices achieve the intended benefits (Jesmin et al. 2012). Other scholars have echoed this call, trying to discern a more nuanced relationship of teamwork as a moderator, a 'complex and adaptive' process that needs to be deployed in the right situations and with the appropriate resources and support to implement these change practices effectively (Belanger and Rodríguez 2008; Bosch et al. 2008; Hann et al. 2007; Wise et al. 2011). There is a dearth of synthesized knowledge about the consideration of implementation factors and the context(s) in which TBPC is most likely to be successful. Thus, focusing on the domains of environment, task and technology—an approach rooted in prior theoretical and empirical work—I explore enabling contextual factors that support the use of TBPC to strengthen primary care delivery.

Approach

This review was informed and structured by adapting the holistic conceptual model developed by the Integrated Team Effectiveness Model (ITEM) (Lemieux-Charles and McGuire 2006). The goal of this review is to determine how a team approach best fits into the current context and dynamic nature of health systems and primary care delivery. Therefore, I explicitly focus on one nation's healthcare system—the USA—and home in on three domains within ITEM that capture the most salient changes likely to be impacting a practice's ability to successfully implement TBPC practices in that system:

1. *Environment* Internal (organizational-level) as well as external (market, policy) characteristics or initiatives that may facilitate or impede TBPC changes;
2. *Task* Specific changes in the scope and nature of health needs and primary care services that may shape the use of TBPC; and
3. *Technology* Currently available technologies that impact practices' ability to effectively implement improved care processes using TBPC.

Whilst these domains do not address the full scope of the ITEM, the implications of research findings in these areas are most likely to be actionable in a policy and practice setting. In addition, though the US-based perspective used in this analysis may limit generalizability, many of the findings summarized below—particularly related to internal organizational culture and teamwork-facilitating structures—may still translate to other nations with different healthcare organization, delivery and financing mechanisms.

Methods

This scoping review of published 'primary care teams' research in the US healthcare setting focuses on how findings in the domains of environment, task and technology better inform our understanding of the enabling contextual factors that promote TBPC. A scoping review was purposely selected because of the applied and dynamic nature of the motivating research question—the need to map available literature and research findings/evidence in a burgeoning area for policymakers and practitioners (Arksey and O'Malley 2005; Kastner et al. 2012).

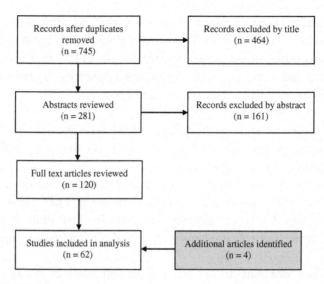

Fig. 5.1 PRISM diagram

Separate searches were conducted in Scopus and PubMed, dating from 2005 to 2015, limited to a more recent time period to capture only the more current environment and contextual factors most salient to providers and policymakers today. I used broad search terms, including 'primary care team' or 'primary care' AND ('teamwork' OR 'team-based'). Articles were restricted to English only.

Figure 5.1 provides a flow diagram for the selection of included studies. Articles were searched and downloaded to an Excel database. I then removed duplicates and reviewed article titles for inclusion, followed by abstract review. Letters, editorials and position statement articles were excluded, as was research that didn't meet the inclusion criteria outlined in Table 5.1. A total of sixty-two articles were included for the full review; these articles were 'charted' and summarized across the three domains of interest (Arksey and O'Malley 2005).

Table 5.1 Scoping review inclusion criteria

Criteria	Description
Consistent definition of primary care team	Research explores the core primary care patient care team, including one primary provider (physician, nurse practitioner or physician's assistant) as well as nurses, assistants, ancillary clinical staff and administration Excludes: Teams that span organizational boundaries Teams focused on inter-physician collaboration Teams not focused on delivery of primary care services Research primarily focused on engaging the patient or caregiver as part of the care team
Focus on organizational context and consideration of implementation factors	Research explores when, why and how practices implemented a team-based care approach to improve care delivery Excludes: Research on the link between teamwork and outcomes, with little or no reflection on why or how teamwork was utilized Research on the interpersonal processes of team formation
Generalizability	Research documenting approaches or strategies that make sense in the US healthcare context Excludes: Research in developing nations lacking a primary care infrastructure Research in other nations that doesn't translate to the US setting Interventions that lack feasibility to promote widely (i.e. substantial, unsustainable influx of temporary staff or resources)

Results

Environment

Applying an 'environmental' lens helps to better understand what organizational and market factors make a setting conducive for deploying a TBPC approach, and where the different levers exist to reshape pressures that may facilitate or impede a 'teams' transition.

Internal Organizational Structures

Research on a 'teams'-focused organizational culture frequently focuses on strong leadership and effective change management as staff deal with the uncertainties and vulnerability of significant role change and altered interpersonal dynamics (Goldman et al. 2010; Hilts et al. 2013). More concretely, practice characteristics that are associated with these enabling organizational strengths include first an organizational philosophy focused on TBPC that is explicitly tied to near-term practice goals and intended changes promoted under PCMH and other delivery reforms (Allan et al. 2014). Physicians' consistent participation in frontline team huddles for daily care planning also sends an important message of physician buy-in and sets an open, collaborative tone for team functioning (Rodriguez et al. 2015a, b). Finally, identifying the personal characteristics in staff that facilitate strong interpersonal dynamics and incorporating them into hiring processes may be an increasingly important organizational strategy to foster high-functioning work relationships amid these changes (Bunniss and Kelly 2008).

Organizations trying to foster teamwork also require appropriate structures and aligned incentives that encourage the effective use of team-based care (TBC) approaches (Hung et al. 2006; Xyrichis and Lowton 2008). Understanding how effectively teams are working together, how the use of teams affects patient outcomes and how to strengthen team functioning all require changes in the traditional ways that practices measure and report performance (Hays 2007; O'Toole et al. 2011). In the context of TBPC, studies have revealed a significant lag between changes in how services are being delivered under new team-based care delivery models and how practice performance is assessed. In an interventional study with 'Family Health Teams' in Ontario, researchers found that existing performance indicators fail to reflect the role and contribution of different team members (Johnston et al. 2011). A qualitative evaluation of sixteen primary care practices in the nationalized healthcare system set up to treat US military veterans (i.e. the US Veterans Health Administration) revealed that performance assessment failed to engage or activate non-physician team members and did little to further the stated organizational goals of emphasizing a team-based care approach (Hysong et al. 2014).

To facilitate enhanced performance initiatives that acknowledge and reinforce team-based care, all team members need access to performance reporting data, both to analyse personal performance and engender a stronger sense of shared responsibility for office functioning and patient care. Performance data should be actionable, with some level of role-specificity (particularly in process measures) to define and maintain individual roles and responsibilities. Having a designated data facilitator driving performance improvement is critical (Watts et al. 2014). Hysong et al. (2014) recommend that—when empowered to do so—a designated nurse or other team member is often better positioned to monitor and manage team processes and outcomes compared to a physician-managed model.

External Policy Environment

Team-based care approaches are typically seen as a facilitator and a mechanism to achieve success under broader delivery reform programmes in the USA, such as accountable care organizations, PCMHs and pay-for-value initiatives (Friedberg et al. 2013; Grace et al. 2014; Grover and Niecko-Najjum 2013). Key elements of a team transformation—(re)negotiating roles, establishing a shared sense of purpose that is patient-centric, fostering open communication etc.—are critical for meeting enhanced practice responsibilities under a PCMH model, particularly in areas such as improved care management and patient engagement/activation (Friedberg et al. 2013; Sanchez and Adorno 2013; True et al. 2014).

Though empirical work is limited, a number of studies have drawn attention to the fact that the financial and regulatory environment of healthcare can seriously impede the implementation of TBC approaches (Finlayson and Raymont 2012). The extent of organizational change possible, especially with respect to enhanced care roles for non-physician staff, is often limited by legal restrictions or ambiguity around role scope. The role of medical assistants in particular is an open question of policy and practical significance as their role continues to expand in the absence of clear regulations, guidelines or best practices (Freund et al. 2015; Ladden et al. 2013).

Existing payment methods in the USA also limit the expansion of TBPC approaches, given the physician-centric fee-for-service billing practices and the task-based nature of reimbursement (McInnes et al. 2015; Strumpf et al. 2012). In a study of salient organizational factors affecting primary care practice in New Zealand, researchers found that degree of inter-office collaboration is strongly influenced by the most prevalent funding mechanism, and that capitation or bulk-funding more strongly promoted the use of teamwork (Pullon et al. 2009). Similar findings from an evaluation of a global payment demonstration in the state of Massachusetts indicated that a transition to use of teams was a critically important component of practices' response to these types of payment reforms (Mechanic et al. 2011).

Task

Primary care providers are faced with an unprecedented workload in today's healthcare environment, coupled with a context of greatly increased documentation and reporting requirements. Whilst shifting to a team-based care approach seems like a natural and often suggested response to these pressures, the ways to best staff and structure these new care teams are less clear. Finlayson and Raymont (2012) emphasize that the 'type, nature and strength of teamwork' is critically shaped by the nature of work itself. Thus, this section explores types of care and service provision that best accommodate a teamwork structure, synthesizing specific findings about how and when teams can be used successfully to accelerate and/or extend care.

Staffing

The development of healthcare teams is often hampered by traditional role concepts (Chesluk and Holmboe 2010). However, increased requirements and resource demands under PCMH and other delivery reform models—particularly around patient-centric care management and coordination—have continued to erode traditional care models and accelerated the development and expansion of new supporting staff roles (Morrissey 2013). Staff in these roles (e.g. care managers, health coaches, navigators etc.) are often best positioned to improve patient activation, connect patients with social services and have the time necessary to manage service utilization and medications; empowered ancillary staff have a particularly well-documented role in the literature in the areas of preventive care and chronic disease management (Altschuler et al. 2012; Edwards et al. 2015; Ferrer et al. 2009; Graffy et al. 2010; Hudson et al. 2007; Margolius et al. 2012). Indeed, a number of recent studies have been able to detect improved patient outcomes as a result of introducing these new care roles (Anand et al. 2010; Chan et al. 2010; Collinsworth et al. 2014). Physicians are busy and have relatively less experience and training in these areas; incorporating new staff to carry out these tasks thus may be viewed as an extension of services rather than an acceleration of existing physician care, and fulfils patient needs complementary to their own role.

Role understanding is a key facilitator in the process of integrating these new staff into the primary care team. Indeed, it is integration rather than collaboration that signifies true embodiment of TBPC principles (Boon et al. 2009). The newness of these care roles, and ambiguity in their defined responsibility and scope, can pose a challenge to physicians and other staff as they incorporate this new team member. Qualitative studies on the role of health coaches emphasize positioning the new staff member as a liaison or as part of a 'relational triad' between the patient and physician. This arrangement cements the coach's role as an advocate and extension of the patients' best interests to improve patient care, and helps physicians and other team members better understand and appreciate this care role (Margolius et al. 2012; Ngo et al. 2010; Wholey et al. 2013). Clear delineation of scope of practice is also critical, as is an explicit understanding of interdependencies. The rest of the team—particularly the physician—needs to understand the functions being performed by the ancillary staff person and how it fits into achieving the broader vision of care that practice aims to provide (Donnelly et al. 2013; Wholey et al. 2013). Finally, securing adequate face-time between 'traditional' and new ancillary team members is critical for fostering an inclusive sense of teamness. Co-location facilitated more acceptance of and reliance on these new care roles, as did including ancillary staff in regular huddles and team meetings (Donnelly et al. 2013; O'Malley et al. 2014).

Structuring

Implementing TBPC requires a difficult navigation of trust, preferences and changed patterns of interaction as roles and responsibilities change, as do organization-level workflows and infrastructure (Mitchell et al. 2012). As new roles take shape, a natural tension and trade-off emerges between role clarity and flexibility. Some argue that a more mechanistic, highly structured team dynamic creates a consistency that builds trust and a feeling of competence (Drach-Zahavy and Freund 2007; Elder et al. 2014). However, a certain degree of flexibility is critical; all team members need to share feelings of responsibility for total patient care that may require completing tasks or services outside his/her defined role description. Clear role guidelines tied to an explicit care mission statement, with guidance on staff cross-coverage expectations, reduce ambiguity to prevent feelings of territorialism or inconsistent TBPC implementation (Grace et al. 2014; Rodriguez et al. 2014).

Relatively discrete clinical tasks (e.g. administering screenings, vaccinations) or administrative tasks (referral tracking, well-visit documentation etc.) can be delegated, as can more nuanced but critical care roles such as patient engagement and education, connecting patients with social services and medication management. Promising strategies to help shape these new enhanced care roles amongst staff include the use of explicit protocols or care templates for routine services and screening (Cross et al. 2015; Goldman et al. 2010; Ladden et al. 2013; O'Malley et al. 2014). Medical assistants (i.e. individuals certified to complete various administrative and low-risk clinical tasks) or ancillary staff trained in case management can often be brought in and trained in panel management and the logistics of care coordination oversight; guidelines and toolkits have been developed to help spread effective role guidelines and best practices for this emerging role (Ladden et al. 2013; Savarimuthu et al. 2013). Making these changes incrementally, with role support through inclusive team huddles and performance feedback, helps build and reinforce these new relationships (O'Malley et al. 2014; Rodriguez et al. 2015a, b).

Technology

Practitioners and researchers have long recognized that the design and functionality of clinical electronic health records (EHRs) shape not just how providers work but also how they can work together (Anand et al. 2010; Bates and Bitton 2010; Howard et al 2012). However, providers lack knowledge on how to use IT to support the holistic changes they are making in pursuit of patient-centric, team-based case (Roper 2014). This is largely due to the underdeveloped state of research exploring the interdependence and synergies of pursuing greater IT implementation in parallel with the use of a team-based approach. There is a dearth of understanding about how care teams learn to work collaboratively within the EHR system, what features facilitate or impede a team approach and how these systems can be designed with new functionality that not only accommodates but also enhances use of a care team.

As primary care teams grow to incorporate new team roles—including care managers, nutritionists, health coaches, etc.—documentation practices need to evolve to support and integrate these new services. Tasks performed and information collected by these ancillary clinical team members is often not incorporated into the central patient record; doctors and nurses often don't see the availability of this information and don't know to act, reiterate or follow-up on this critical resource (Cross et al. 2015; Donnelly et al. 2013; Kim et al. 2013; O'Malley et al. 2015). Available documentation features also tend to be 'flat', lacking some of the advanced features such as branching logic and decision support to act on collected information and provide enhanced management, education and support services to patients.

Developing EHRs that support a team-based workflow requires an intimate understanding of how team members work together and interface with technology support and documentation practices. Team members need to be able to complete tasks but also communicate about patient care and rely on EHR-facilitated reminders and workflow support to track and ensure the follow-up/reconciliation of pending responsibilities. In one of the few studies to explore the interaction of team behaviour and effective EHR use, authors identified the importance of team agreement on methods of communication and the consistency of EHR role and documentation practices (Denomme et al. 2011). For EHRs to function as a reliable coordination platform for patient care, all team members need to know where specific information should be recorded and can be retrieved; tracking and other automated decision support or registry functionality also requires consistent (and complete) documentation. Other studies have mentioned the availability of a limited set of internal communication and coordination tools within the EHR (Cross et al. 2015; Donnelly et al. 2013; Legault et al. 2012; O'Malley et al. 2015). However, this has yet to be the subject of rigorous exploration or optimization.

Technology can also be deployed to facilitate the expansion of team member responsibilities and autonomy beyond traditional roles. Existing studies mention team-based care approaches in concert with the use of care templates (Cross et al. 2015; Graffy et al. 2010; Kendall et al. 2013; O'Malley et al. 2015), panel management tools (Kaferle and Wimsatt 2012), registries (Graffy et al. 2010; Rodriguez et al. 2014) and the use of patient engagement tools/shared decision-making applications (Chunchu et al. 2012; Friedberg et al. 2013), yet stop short of identifying synergies in these concepts. More conceptual work and empirical analyses remain to be done to understand the full implications of integrating these technologies with a TBC practice design, including the specific challenges and potential legal or financial ramifications of using IT to support and extend patient care roles for non-physician team members.

Discussion

This review synthesizes available research exploring how three key contextual factors—environment, task and technology—shape primary care practices' implementation of team-based approaches to primary care. Important environmental considerations that emerged from existing literature include strong and invested physician leadership, performance measurement practices that reflect and support a 'teams' approach and reimbursement structures that facilitate enhanced use of non-physician staff. The changing nature of tasks and workflow in primary care service delivery, including an elevated focus on preventive care, patient engagement and disease management, bring into focus a need for new staffing models and an efficient restructuring of roles to support new care practices. Technology applications (e.g. EHRs, registries etc.) can support and enhance team-based care practices by enhancing communication, coordination, role support and care quality assurance.

Practice-level efforts to implement TBPC practices may involve significant restructuring of physician and non-physician team roles to survive in a changing healthcare environment. Physicians and administrators need to spearhead changes in organizational culture to support this level of learning and change. This includes clear goal-setting and commitment to supporting new team-based models through changes in compensation, physical infrastructure and how practice performance is measured, evaluated and acted upon. Educational programmes and interventions to facilitate a 'teams transformation' may prove useful, distilling key principles and tools to help with the interpersonal, psychosocial processes of developing well-functioning teams (Chan et al. 2010; Kozlowski and Ilgen 2006).

A second key implementation factor to consider is that the adoption of team-based care practices doesn't take place in a vacuum. TBPC efforts can both support but also be influenced by other concurrent practice changes promoted under PCMH, such as enhanced patient engagement and patient-centric care management services as well as the use of EHRs. Information technology can help to support structured efforts to enhance non-physician care team roles, improve coordination and workflow and facilitate communication amongst team members. However, a number of continued challenges hinder practices' ability to effectively leverage these strategies. For example, practices need to develop procedures to deal with asynchronous communication within the team, and figure out how to integrate documentation practices across multiple care team providers in a systematic way that makes finding and sharing patient information easy and reliable.

At the state and national policy level, reimbursement practices need to continue to shift away from a physician-based fee-for-service model and acknowledge new care team practice models. This includes a focus on pay-for-value, but more broadly requires acknowledging the care roles of non-physician staff in task-based reimbursement. Without aligned financial incentives, practice physicians and administrators will find it much harder to sustain TBPC efforts. Any efforts to reform reimbursement structure will also require clearer guidelines on the education, role and scope of practice for non-physician care team members. This is particularly true in the case of medical assistants, whose numbers continue to grow exponentially and whose role varies widely across practices.

Conclusion

This study is the first to synthesize available research on the contextual factors that impact the implementation of team-based care practices in primary care settings. Focusing on three key domains of environment, task and technology, I explore the conditions under which practices can most effectively leverage TBPC strategies, and identify key ways to foster more effective TBPC in the future. These findings enhance our understanding of the tenuous link between the adoption of TBPC and the improvement of patient as well as practice-level outcomes. Further research should consider the findings of this review to improve the nuance of empirical analyses in this area, seeking to explain not just *whether* using teams works, but *when*. Studying the mechanisms through which the adoption of team-based care practices can lead to better outcomes—both mediating structures or processes such as the consistent use of huddles, as well as moderating organizational factors like presence of an EHR or participation in pay-for-value programmes—provide actionable findings to improve policy reforms, organizational change processes and ultimately patient care.

References

Allan, H. T., Brearley, S., Byng, R., Christian, S., Clayton, J., MacKintosh, M., & Ross, F. (2014). People and teams matter in organizational change: Professionals' and managers' experiences of changing governance and incentives in primary care. *Health Services Research, 49*(1), 93–112.

Altschuler, J., Margolius, D., Bodenheimer, T., & Grumbach, K. (2012). Estimating a reasonable patient panel size for primary care physicians with team-based task delegation. *Annals of Family Medicine, 10*(5), 396–400.

Anand, S. G., Adams, W. G., & Zuckerman, B. S. (2010). Specialized care of overweight children in community health centers. *Health Affairs, 29*(4), 712–717.

Arksey, H., & O'Malley, L. (2005). Scoping studies: Towards a methodological framework. *International Journal of Social Research Methodology, 8*(1), 19–32.

Bates, D. W., & Bitton, A. (2010). The future of health information technology in the patient-centered medical home. *Health Affairs, 29*(4), 614–621.

Belanger, E., & Rodríguez, C. (2008). More than the sum of its parts? A qualitative research synthesis on multi-disciplinary primary care teams. *Journal of Interprofessional Care, 22*(6), 587–597.

Boon, H. S., Mior, S. A., Barnsley, J., Ashbury, F. D., & Haig, R. (2009). The difference between Integration and collaboration in patient care: Results from key informant interviews working in multiprofessional health care teams. *Journal of Manipulative and Physiological Therapeutics, 32*(9), 715–722.

Bosch, M., Dijkstra, R., Wensing, M., Van Der Weijden, T., & Grol, R. (2008). Organizational culture, team climate and diabetes care in small office-based practices. *BMC Health Services Research, 8*(1), 180.

Bunniss, S., & Kelly, D. R. (2008). 'The unknown becomes the known': Collective learning and change in primary care teams. *Medical Education, 42*(12), 1185–1194.

Chan, B. C., Perkins, D., Wan, Q., Zwar, N., Daniel, C., Crookes, P., & Harris, M. F. (2010). Finding common ground? Evaluating an intervention to improve teamwork among primary health-care professionals. *International Journal for Quality in Health Care, 22*(6), 519–524.

Chesluk, B. J., & Holmboe, E. S. (2010). How teams work—Or don't—In primary care: A field study on internal medicine practices. *Health Affairs, 29*(5), 874–879.

Chunchu, K., Mauksch, L., Charles, C., Ross, V., & Pauwels, J. (2012). A patient centered care plan in the EHR: Improving collaboration and engagement. *Families, Systems, & Health, 30*(3), 199.

Collinsworth, A., Vulimiri, M., Snead, C., & Walton, J. (2014). Community health workers in primary care practice redesigning health care delivery systems to extend and improve diabetes care in underserved populations. *Health Promotion Practice, 15*(Suppl. 2), 51S–61S.

Cross, D. A., Cohen, G. R., Nong, P., Day, A.-V., Vibbert, D., Naraharisetti, R., & Adler-Milstein, J. (2015). Improving EHR capabilities to facilitate stage 3 meaningful use care coordination criteria. *AMIA Annual Symposium Proceedings, 5*, 448–455.

Cutrona, S. L., Choudhry, N. K., Fischer, M. A., Servi, A., Liberman, J. N., Brennan, T., & Shrank, W. H. (2010). Modes of delivery for interventions to improve cardiovascular medication adherence: Review. *American Journal of Managed Care, 16*(12), 929.

Denomme, L. B., Terry, A. L., Brown, J. B., Thind, A., & Stewart, M. (2011). Primary health care teams' experience of electronic medical record use after adoption. *Family Medicine, 43*(9), 632–642.

Donaldson, M. S., Yordy, K. D., Lohr, K. N., & Vanselow, N. A. (1996). *Primary care: America's health in a new era.* Washington, DC: National Academies Press.

Donnelly, C., Brenchley, C., Crawford, C., & Letts, L. (2013). The integration of occupational therapy into primary care: A multiple case study design. *BMC Family Practice, 14*(1), 60.

Drach-Zahavy, A., & Freund, A. (2007). Team effectiveness under stress: A structural contingency approach. *Journal of Organizational Behavior, 28*(4), 423–450.

Edwards, S. T., Rubenstein, L. V., Meredith, L. S., Hackbarth, N. S., Stockdale, S. E., Cordasco, K. M., & Yano, E. M. (2015). Who is responsible for what tasks within primary care: Perceived task allocation among primary care providers and interdisciplinary team members. *Healthcare, 3*(3), 142–149.

Elder, N. C., Jacobson, C. J., Bolon, S. K., Fixler, J., Pallerla, H., Busick, C., ... Pugnale, M. (2014). Patterns of relating between physicians and medical assistants in small family medicine offices. *The Annals of Family Medicine, 12*(2), 150–157.

Ferrer, R. L., Mody-Bailey, P., Jaén, C. R., Gott, S., & Araujo, S. (2009). A medical assistant-based program to promote healthy behaviors in primary care. *The Annals of Family Medicine, 7*(6), 504–512.

Finlayson, M. P., & Raymont, A. (2012). Teamwork-general practitioners and practice nurses working together in New Zealand. *Journal of Primary Health Care, 4*(2), 150–155.

Freund, T., Everett, C., Griffiths, P., Hudon, C., Naccarella, L., & Laurant, M. (2015). Skill mix, roles and remuneration in the primary care workforce: Who are the healthcare professionals in the primary care teams across the world? *International Journal of Nursing Studies, 52*(3), 727–743.

Friedberg, M. W., Van Busum, K., Wexler, R., Bowen, M., & Schneider, E. C. (2013). A demonstration of shared decision making in primary care highlights barriers to adoption and potential remedies. *Health Affairs, 32*(2), 268–275.

Friedman, E. L., Chawla, N., Morris, P. T., Castro, K. M., Carrigan, A. C., Das, I. P., & Clauser, S. B. (2014). Assessing the development of multidisciplinary care: Experience of the National Cancer Institute Community Cancer Centers Program. *Journal of Oncology Practice, 11*(1), 36–43.

Goldman, J., Meuser, J., Lawrie, L., Rogers, J., & Reeves, S. (2010). Interprofessional primary care protocols: A strategy to promote an evidence-based approach to teamwork and the delivery of care. *Journal of Interprofessional Care, 24*(6), 653–665.

Grace, S. M., Rich, J., Chin, W., & Rodriguez, H. P. (2014). Flexible implementation and integration of new team members to support patient-centered care. *Healthcare, 2*(2), 145–151.

Graffy, J., Grant, J., Williams, K., Cohn, S., Macbay, S., Griffin, S., & Kinmonth, A. L. (2010). More than measurement: Practice team experiences of screening for type 2 diabetes. *Family Practice, 27*(4), 386–394.

Grover, A., & Niecko-Najjum, L. M. (2013). Primary care teams: Are we there yet? Implications for workforce planning. *Academic Medicine, 88*(12), 1827–1829.

Hann, M., Bower, P., Campbell, S., Marshall, M., & Reeves, D. (2007). The association between culture, climate and quality of care in primary health care teams. *Family Practice, 24*(4), 323–329.

Hays, R. (2007). Measuring quality in the new era of team-based primary care. *Quality in Primary Care, 15*(3), 133–135.

Helfrich, C. D., Dolan, E. D., Simonetti, J., Reid, R. J., Joos, S., Wakefield, B. J. and Nelson, K. (2014). Elements of team-based care in a patient-centered medical home are associated with lower burnout among VA primary care employees. *Journal of General Internal Medicine, 29*(Suppl. 2), S659–S666.

Hilts, L., Howard, M., Price, D., Risdon, C., Agarwal, G., & Childs, A. (2013). Helping primary care teams emerge through a quality improvement program. *Family Practice, 30*(2), 204–211.

Howard, J., Shaw, E. K., Felsen, C. B., & Crabtree, B. F. (2012). Physicians as inclusive leaders: Insights from a participatory quality improvement intervention. *Quality Management in Health Care, 21*(3), 135–145.

Hudson, S. V., Ohman-Strickland, P., Cunningham, R., Ferrante, J. M., Hahn, K., & Crabtree, B. F. (2007). The effects of teamwork and system support on colorectal cancer screening in primary care practices. *Cancer Detection and Prevention, 31*(5), 417–423.

Hung, D. Y., Rundall, T. G., Crabtree, B. F., Tallia, A. F., Cohen, D. J., & Halpin, H. A. (2006). Influence of primary care practice and provider attributes on preventive service delivery. *American Journal of Preventive Medicine, 30*(5), 413–422.

Hysong, S. J., Knox, M. K., & Haidet, P. (2014). Examining clinical performance feedback in patient-aligned care teams. *Journal of General Internal Medicine, 29*(Suppl. 2), S667–S674.

Jesmin, S., Thind, A., & Sarma, S. (2012). Does team-based primary health care improve patients' perception of outcomes? Evidence from the 2007–2008 Canadian survey of experiences with primary health. *Health Policy, 105*(1), 71–83.

Johnston, S., Green, M., Thille, P., Savage, C., Roberts, L., Russell, G., & Hogg, W. (2011). Performance feedback: An exploratory study to examine the acceptability and impact for interdisciplinary primary care teams. *BMC Family Practice, 12*(1), 14.

Kaferle, J. E., & Wimsatt, L. A. (2012). A team-based approach to providing asthma action plans. *Journal of the American Board of Family Medicine, 25*(2), 247–249.

Kastner, M., Tricco, A. C., Soobiah, C., Lillie, E., Perrier, L., Horsley, T., ... Straus, S. E. (2012). What is the most appropriate knowledge synthesis method to conduct a review? Protocol for a scoping review. *BMC Medical Research Methodology, 12*(1), 114.

Kendall, M., Mason, B., Momen, N., Barclay, S., Munday, D., Lovick, R., ... Cormie, P. (2013). Proactive cancer care in primary care: A mixed-methods study. *Family Practice, 30*(3), 302–312.

Kim, T. W., Saitz, R., Kretsch, N., Cruz, A., Winter, M. R., Shanahan, C. W., & Alford, D. P. (2013). Screening for unhealthy alcohol and other drug use by health educators: Do primary care clinicians document screening results? *Journal of Addiction Medicine, 7*(3), 204–209.

Kozlowski, S. W., & Ilgen, D. R. (2006). Enhancing the effectiveness of work groups and teams. *Psychological Science in the Public Interest, 7*(3), 77–124.

Ladden, M. D., Bodenheimer, T., Fishman, N. W., Flinter, M., Hsu, C., Parchman, M., & Wagner, E. H. (2013). The emerging primary care workforce: Preliminary observations from the primary care team: Learning from effective ambulatory practices project. *Academic Medicine, 88*(12), 1830–1834.

Legault, F., Humbert, J., Amos, S., Hogg, W., Ward, N., Dahrouge, S., & Ziebell, L. (2012). Difficulties encountered in collaborative care: Logistics trumps desire. *Journal of the American Board of Family Medicine, 25*(2), 168–176.

Lemieux-Charles, L., & McGuire, W. L. (2006). What do we know about health care team effectiveness? A review of the literature. *Medical Care Research and Review, 63*(3), 263–300.

Margolius, D., Wong, J., Goldman, M. L., Rouse-Iniguez, J., & Bodenheimer, T. (2012). Delegating responsibility from clinicians to nonprofessional personnel: The example of hypertension control. *Journal of the American Board of Family Medicine, 25*(2), 209–215.

McAllister, J. W., Carl Cooley, W., Van Cleave, J., Boudreau, A. A., & Kuhlthau, K. (2013). Medical home transformation in pediatric primary care-what drives change? *Annals of Family Medicine, 11*(SUPPL. 1), S90–S98.

McInnes, S., Peters, K., Bonney, A., & Halcomb, E. (2015). An integrative review of facilitators and barriers influencing collaboration and teamwork between general practitioners and nurses working in general practice. *Journal of Advanced Nursing, 71*(9), 1973–1985.

Mechanic, R. E., Santos, P., Landon, B. E., & Chernew, M. E. (2011). Medical group responses to global payment: Early lessons from the 'alternative quality contract' in Massachusetts. *Health Affairs, 30*(9), 1734–1742.

Mitchell, P., Wynia, M., Golden, R., McNellis, B., Okun, S., Webb, C. E., Von Kohorn, I. (2012). *Core principles & values of effective team-based health care.* Washington, DC: Institute of Medicine.

Mohr, D. C., Benzer, J. K., & Young, G. J. (2013). Provider workload and quality of care in primary care settings: Moderating role of relational climate. *Medical Care, 51*(1), 108–114.

Morrissey, J. (2013). Filling the gaps in primary care: New roles strengthen relationships with patients. *Trustee: The Journal for Hospital Governing Boards, 66*(4), 8–10.

Ngo, V., Hammer, H., & Bodenheimer, T. (2010). Health coaching in the teamlet model: A case study. *Journal of General Internal Medicine, 25*(12), 1375–1378.

O'Malley, A. S., Gourevitch, R., Draper, K., Bond, A., & Tirodkar, M. A. (2014). Overcoming challenges to teamwork in patient-centered medical homes: A qualitative study. *Journal of General Internal Medicine, 30*(2), 183–192.

O'Malley, A. S., Draper, K., Gourevitch, R., Cross, D. A., & Scholle, S. H. (2015). Electronic health records and support for primary care teamwork. *Journal of the American Medical Informatics Association, 22*(2), 426–434.

O'Toole, T. P., Cabral, R., Blumen, J. M., & Blake, D. A. (2011). Building high functioning clinical teams through quality improvement initiatives. *Quality in Primary Care, 19*(1), 13–22.

Page, N. (2006). Task overlap among primary care team members: An opportunity for system redesign? *Journal of Healthcare Management, 51*(5), 295.

Pullon, S., McKinlay, E., & Dew, K. (2009). Primary health care in New Zealand: The impact of organisational factors on teamwork. *British Journal of General Practice, 59*(560), 191–197.

Roblin, D. W., Howard, D. H., Junling, R., & Becker, E. R. (2011). An evaluation of the influence of primary care team functioning on the health of Medicare beneficiaries. *Medical Care Research & Review, 68*(2), 177–201.

Rodriguez, H. P., Chen, X., Martinez, A. E., & Friedberg, M. W. (2015a). Availability of primary care team members can improve teamwork and readiness for change. *Health Care Management Review, 41*(4), 286–295.

Rodriguez, H. P., Ivey, S. L., Raffetto, B. J., Vaughn, J., Knox, M., Hanley, H. R., ... Shortell, S. M. (2014). As good as it gets? Managing risks of cardiovascular disease in California's top-performing physician organizations. *Joint Commission Journal on Quality and Patient Safety, 40*(4), 148–158.

Rodriguez, H. P., Meredith, L. S., Hamilton, A. B., Yano, E. M., & Rubenstein, L. V. (2015b). Huddle up! The adoption and use of structured team communication for VA medical home implementation. *Health Care Management Review, 40*(4), 286–299.

Roper, R. (2014). Insights on innovative strategies for harnessing health information technology to help individuals, teams. *Journal of Ambulatory Care Management, 37*(2), 96–99.

Saba, G. W., Villela, T. J., Chen, E., Hammer, H., & Bodenheimer, T. (2012). The myth of the lone physician: Toward a collaborative alternative. *The Annals of Family Medicine, 10*(2), 169–173.

Sanchez, K., & Adorno, G. (2013). "It's like being a well-loved child": Reflections from a collaborative care team. *The Primary Care Companion for CNS Disorders, 15*(6), PCC.13m01541. http://doi.org/10.4088/PCC.13m01541.

Savarimuthu, S. M., Jensen, A. E., Schoenthaler, A., Dembitzer, A., Tenner, C., Gillespie, C., ... Sherman, S. E. (2013). Developing a toolkit for panel management: Improving hypertension and smoking cessation outcomes in primary care at the VA. *BMC Family Practice, 14*(1), 176.

Shipman, S. A., & Sinsky, C. A. (2013). Expanding primary care capacity by reducing waste and improving the efficiency of care. *Health Affairs, 32*(11), 1990–1997.

Starfield, B., Shi, L., & Macinko, J. (2005). Contribution of primary care to health systems and health. *Milbank Quarterly, 83*(3), 457–502.

Strumpf, E., Levesque, J.-F., Coyle, N., Hutchison, B., Barnes, M., & Wedel, R. J. (2012). Innovative and diverse strategies toward primary health care reform: lessons learned from the Canadian experience. *Journal of the American Board of Family Medicine, 25*(Suppl 1), S27–S33.

Thomas, M. H. (2014). Development and implementation of a pharmacist-delivered Medicare annual wellness visit at a family practice office. *Journal of the American Pharmacists Association, 54*(4), 427–434.

True, G., Stewart, G. L., Lampman, M., Pelak, M., & Solimeo, S. L. (2014). Teamwork and delegation in medical homes: Primary care staff perspectives in the Veterans Health Administration. *Journal of General Internal Medicine, 29*(Suppl. 2), S632–S639.

Watts, B., Lawrence, R. H., Singh, S., Wagner, C., Augustine, S., & Singh, M. K. (2014). Implementation of quality improvement skills by primary care teams: Case study of a large academic practice. *Journal of Primary Care and Community Health, 5*(2), 101–106.

Wholey, D. R., White, K. M., Adair, R., Christianson, J. B., Lee, S., & Elumba, D. (2013). Care guides: An examination of occupational conflict and role relationships in primary care. *Health Care Management Review, 38*(4), 272–283.

Willard-Grace, R., Hessler, D., Rogers, E., Dubé, K., Bodenheimer, T., & Grumbach, K. (2014). Team structure and culture are associated with lower burnout in primary care. *Journal of the American Board of Family Medicine, 27*(2), 229–238.

Wise, C. G., Alexander, J. A., Green, L. A., Cohen, G. R., & Koster, C. R. (2011). Journey toward a patient-centered medical home: Readiness for change in primary care practices. *Milbank Quarterly, 89*(3), 399–424.

Xyrichis, A., & Lowton, K. (2008). What fosters or prevents interprofessional teamworking in primary and community care? A literature review. *International Journal of Nursing Studies, 45*(1), 140–153.

6

Doing More with Less: *Lean Healthcare* Implementation in Irish Hospitals

Mary A. Keating and Brendan S. Heck

Introduction

Worldwide, there is pressure on public services to become more efficient. For healthcare, this includes addressing challenges associated with ageing populations and chronic diseases at a time of resource constraint. Healthcare organisations need to deliver quality care and extend service levels whilst simultaneously controlling expenditure (Waring and Bishop 2010). Since the early 2000s *Lean*—a well-known service improvement approach—has been adopted to reconcile and achieve these goals (Brandao de Souza 2009; D'Andreamatteo et al. 2015). Reflecting this, *Lean* is emerging as a key component in the literature concerning service improvements in health systems.

Inspired by the work of Burgess and Radnor (2013), who examined the status of *Lean* implementation in hospitals in the English National

M.A. Keating (✉) · B.S. Heck
Trinity Business School, Trinity College Dublin, Dublin, Ireland
e-mail: mkeating@tcd.ie

© The Author(s) 2018
A.M. McDermott et al. (eds.), *Managing Improvement in Healthcare*,
Organizational Behaviour in Health Care,
https://doi.org/10.1007/978-3-319-62235-4_6

Health Service (NHS), this chapter aims to determine how *Lean* is being applied in the Irish healthcare sector. *Lean* initiatives, especially focused around nursing practice, have been tested in a number of hospitals in Ireland (White et al. 2014). However, the overall situation regarding the implementation of *Lean* in Irish hospitals is an area that merits further investigation. Concretely, the following research questions are addressed:

- How are *Lean* methods and processes implemented in Irish public hospitals?
- What factors are influencing *Lean healthcare* implementation in Ireland?
- How does the Irish approach to implementing *Lean healthcare* compare when set against the approach to *Lean* reported for other countries?

This chapter starts by defining *Lean*, before outlining the current evidence regarding *Lean* in healthcare. The qualitative method is then detailed, before a presentation of the findings of this empirical pilot study. Finally, conclusions regarding the current state of *Lean* implementation in the Irish healthcare context, and its contribution towards creating, spreading and sustaining improvement in the Irish context, are outlined.

Background

Defining *Lean* is not straightforward (Pettersen 2009). An improvement philosophy, it originated at the Toyota Corporation in Japan in the 1950s and is sometimes referred to as the *Toyota Production System* or *TPS*. *Lean* thinking focuses on customer value, defined as the ability to deliver the exact product or service required by customers in a timely manner and at an appropriate price. It is premised on five key operational principles (Womack and Jones 1996):

- Value—Specifying the value desired by customers;
- Value stream—Identifying the value stream for each product, providing value and challenging all wasted steps;

- Flow—Making the product or system flow continuously;
- Pull—Introducing pull between all steps where continuous flow is impossible;
- Perfection—Managing towards perfection in order to reduce the amount of time and the amount of steps needed to serve the customer.

Recent literature suggests that *Lean* should be considered as a systemic quality improvement approach and not simply as a set of specific tools that enable improvements (Burgess 2012). We also note that discussion of *Lean* often incorporates *Six Sigma*, and it can be referred to as *Lean Six Sigma (LSS)* (Shah et al. 2008). Both are considered to be 'process improvement programmes', described as 'synergistic' (Bossert 2003: 31) as they are similar in approach to systemic quality management (Proudlove et al. 2008).

In healthcare, *Lean* is considered a strategic approach which enables hospitals to reduce delays and errors whilst improving the quality of care through involving their staff in a process of continuous improvement (Graban 2008). Toussaint and Gerard (2010) summarised *Lean* principles for the healthcare context in three points: focusing on and designing care around patients; identifying value for patients and eliminating waste; and minimising the waiting time for treatment as well as treatment time. Much of the benefit from *Lean* for healthcare organisations derives from its promotion of more efficient processes. This may enable savings to be made whilst also providing higher quality care, thereby promoting better value for patients. Additional positive outcomes derived from *Lean* include improved access, efficiency and quality of medical care as well as reduced mortality, whilst the empowerment of employees, the introduction of gradual continuous improvement and the resulting increase in accountability can be considered as further beneficial aspects (Mazzocato et al. 2010).

There is evidence of successful application of *Lean* to achieve these outcomes in health services around the world, including in the most prevalent adopters—the USA and the UK (Brandao de Souza 2009; D'Andreamatteo et al. 2015). Yet despite this potential, its application has been described as narrow, piecemeal and disjointed, characterised by the application of specific *Lean* tools in distinct quality improvement projects or programmes (Poksinska 2010). It seems that *Lean* is not being implemented using the holistic and integrated approach advocated in the literature summarised by Burgess (2012, p. 65), who notes that '[t]he extant literature makes a very clear case that *Lean* as derived from the TPS should be understood as a holistic approach to continuous improvement and not a set of tools'. Sustainable results appear to be dependent on creating a change culture involving a longer term vision of continuous improvement (Radnor and Osborne 2013). It may be that because *Lean healthcare* is a relatively new field, its implementation is still at an early stage of development. Alternatively, barriers may negatively affect its prospects. These are considered below.

Brandao de Souza and Pidd (2011) identify major barriers to *Lean* implementation in healthcare settings. Some of these are unique to healthcare. Key barriers include professional and functional silos, hierarchy and resistance to change. In addition, failure to achieve readiness factors, such as leadership, training, organisational culture and communication (Al-Balushi et al. 2014) may be an impediment. Overall, *Lean* appears to bring about positive results when applied in a healthcare setting, but researchers have identified limitations which prevent general conclusions from being drawn regarding its overall impact (D'Andreamatteo et al. 2015).

Based on this succinct summary, one might expect a patchy implementation of *Lean* in Irish hospitals. The method by which this is explored is detailed in the next section.

Methodology

This part-replication study set out to investigate how *Lean* is being applied in Irish hospitals. Following Burgess and Radnor (2013), it combined content analysis of hospital annual reports with additional narrative analysis of interviews with recognised Irish *Lean healthcare* experts. The research objectives were to identify how *Lean* is being implemented in Irish hospitals and to apply the Lean implementation classification developed by Burgess (2012), to establish how the *Lean healthcare* process implementation is being carried out.

Phase 1—Secondary Source Data Collection and Analysis Within Irish Hospitals

The Irish health sector was undergoing significant restructuring at the time of data collection with the regrouping of fifty separate hospitals into seven distinct Hospital Groups (Health Service Executive 2015). Therefore, it was decided to focus on the seven main, large, multidisciplinary acute hospitals in Ireland for the purpose of this exploratory study. In this phase of the research, a content analysis of recently published annual reports (2013) was carried out, using a combined 'key word in context' and 'narrative analysis' approach as described by Grbich (2007). The three dimensions of Pettigrew and Whipp's (1991) Context-Content-Process model of strategic change adapted by Burgess and Radnor (2013) informed this content analysis. These three dimensions refer to the 'why', the 'what' and the 'how' of change.

Phase 2—Qualitative Interviews with Experts

The second phase of this research involved a narrative analysis of qualitative interview data. The purpose of these interviews was to contextualise the findings and facilitate a better analysis. Three semi-structured interviews were undertaken with prominent experts on *Lean healthcare* in Ireland: two certified *LSS* Black Belts both widely recognised as highly competent in the *LSS methodology* and leading quality improvement projects in a full-time capacity in Ireland, and an expert who has written about specific aspects of *Lean* implementation.

Findings

Annual report statements from the Chairperson and/or from the chief executive officer (CEO) of each hospital provided a narrative and offered valuable insight into the strategic context, processes and content of *Lean* and/or *LSS* implementation in the sample of Irish hospitals. Based on this content analysis, the following key words and the rationale for selecting them were identified. These were judged to be linked with the implementation of *Lean healthcare* and/or *LSS*; some are identical to those used by Burgess and Radnor (2013):

- 'innovation'—referring to introducing new processes and projects which may involve *Lean* and/or *LSS*;
- 'reconfiguration'—linked to reorganisation and merging which may demonstrate that *Lean* and/or *LSS* methods are being implemented.
- 'pathways'—referring to patient pathways and the improvement of patient flow within them which is associated with *Lean* and/or *LSS*;
- 'value'—referring to identifying, specifying and increasing the value for patients;
- 'lean'—referring to knowledge or application of *Lean* and/or *LSS* approaches and methodologies;
- 'integrat'—base form of word integration which may describe processes of standardisation and improvement of systems, including clinical and information technology systems, commonly linked to *Lean* and/or *LSS*;
- 'waste'—referring to removing of waste in processes;
- 'quality', 'safety' and 'improvement' or 'QSI'—referring to process improvement initiatives and programmes which may be associated with *Lean* and/or *LSS*;
- 'improvement'—activities linked to quality improvement or service improvement which may indicate *Lean* and/or *LSS*;
- 'optimis'—base form of word optimising, synonymous with improving;
- 'initiatives'—synonymous with project and can identify initiatives associated with *Lean* and/or *LSS* methods;
- 'project'—identifying various projects which may involve *Lean* and/or *LSS* methods;
- 'productive'—referring to the implementation of the Productive Ward (PW) programme which is associated with *Lean* implementation;
- 'strateg'—base form of the word strategy, which may denote a strategic shift using process and quality improvements associated with *Lean* and/or *LSS*;

- 'process'—referring to process improvement which is intrinsically linked with *Lean* and/or *LSS*;
- 'performance'—referring to performance optimisation through continuous improvement which is associated with *Lean* and/or *LSS*;
- 'staff'—referring to staff cooperation and staff buy-in which are intrinsically linked to *Lean* and/or *LSS* implementation.

The *Lean* implementation classification developed and described by Burgess (2012) and Burgess and Radnor (2013) contained five categories. This was modified marginally through the introduction of a sixth category, 'No *Lean*', and used to guide the analysis of the content data. Therefore, the six categories of approaches to *Lean* implementation presented and described below were used to categorise the key words and determine the approach of Irish hospitals to the implementation of *Lean*:

- No *Lean*—no indication of *Lean* found[1];
- Tentative—the hospital is contemplating *Lean*; tendering for external management consultancy to help with implementation or piloting a small isolated project;
- Productive Ward (PW) only—the hospital is implementing Productive Ward and/or Productive Theatre but no other evidence of *Lean* implementation is identified;
- Few projects—the hospital is using *Lean* principles and methods to underpin projects relating to certain functions or pathways within the organisation;
- Programme—the hospital managers refer to *Lean* principles underpinning work programmes expected to last between one and five years;
- Systemic—the hospital reports refer to embedding *Lean* principles in the hospital as a whole so that it becomes the standard. A systemic implementation also emphasises *Lean* training for all staff.

---

[1]A minor modification involved adding a 'No *Lean*' category and replacing the word 'trust' by the word 'hospital' in order to ensure relevance to the Irish healthcare system.

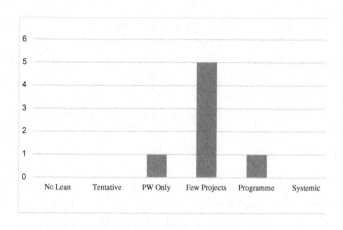

Fig. 6.1 Lean implementation in Irish acute hospitals

Lean application varied from 'PW only' to a 'programme' approach. It appears that having a 'few projects' was the approach to *Lean* implementation most common in Irish hospitals with five of the seven hospitals being classified as such. Figure 6.1 presents the overall findings from Phase 1 and presents a snapshot illustrating the distribution of the approaches to *Lean* implementation at the relevant point in time.

The results were as anticipated, with no hospital adopting a systemic approach and the dominant approach being one of implementing a 'few projects'.

Next, we detail the findings from the Phase 2 qualitative interviews. The interviews were conducted with influential stakeholders and practitioners in the area of *Lean* working in the Irish healthcare system. Their views provide a broader insight into *Lean healthcare* implementation in the context of service quality improvement in Ireland and serve to contextualise and support the analysis of the *Lean healthcare* implementation snapshot provided by Phase 1. Interviews were recorded in person, transcribed and analysed through narrative analysis. Interviewee A, an *LSS* Black Belt practitioner working in a large urban hospital, explained that *Lean* had become part of the philosophy and strategy of the hospital, stating that:

> Our goal is to be the first Lean hospital in Ireland and our second goal is to be the first Lean hospital group. The goal from the outset has been to

create a Lean culture as part of the transformational change within the hospital.

In Interviewee A's opinion, *Lean* in isolation does not work as it is part of the total service improvement process in the hospital, depending on and complementing other quality improvement initiatives. The hospital is moving towards a 'systemic' *Lean* implementation approach where *Lean* will become the standard across all hospital services. Interviewee A suggests that leadership and buy-in from all staff—medical and administrative—are equally important for successful *Lean* implementation. He warns that a 'toolbox' approach to implementing *Lean* could fail if staff are not provided with appropriate training. In this hospital, the *Lean* training model is inspired by best practice in the USA, the UK and Australia, but tailored to the needs of the hospital.

Interviewee A explained that *Lean* has a role to play in addressing 'silos in healthcare', enabling effective team integration in providing patient-centred services, but he stated that 'islands of best practice' can also create 'silos of *Lean*' within healthcare organisations. He provided evidence that in the hospital indirect financial benefits have been derived from *Lean*. Based on his previous experience of working with *Lean* implementation in hospitals in Ireland and abroad, he stated that a single approach to *Lean* may not work in all hospitals and that the implementation context needs to be taken into consideration.

Interviewee B works as a service quality improvement champion in the Irish Health Service Executive (HSE)—the national body responsible for the provision of health and personal care services in Ireland. Interviewee B stated that service improvement in healthcare is generally not dependant on a specific quality improvement process or tool, and that the Irish HSE's recommended approach to it can be considered an 'eclectic mix'. *Lean* is simply one approach that can be adopted in the Irish system. He described the take-up of *Lean* across the Irish hospital sector as 'sporadic' with 'specific islands of improvement'. Reflecting on why this was the case, he commented that the turnover of senior management in Irish hospitals may be contributing to the relatively conservative approach and slow take-up of initiatives such as *Lean*. Where *Lean* has been implemented, hospital managerial leadership has been a critical influencing factor, supported by an emphasis on staff engagement in the process and training to enable successful implementation.

Interviewee C provides technical support and advice to hospitals interested in implementing *Lean* processes. He commented that continuous improvement is very challenging as well as complex in health service organisations, and that understanding the impact and benefits of the change process from the patients as well as the service provider's perspective is important. Interviewee C stressed that putting the patient at the centre of the improvement process could bring about safe quality care as well as streamlining processes. He asserted that in implementing a Lean improvement process, it is important to adopt an organisation-wide perspective and to work on specific improvement projects which complement each other. For successful implementation of a quality change project such as *Lean*, Interviewee C concurred with previous interviewees that managerial leadership, staff buy-in and training were important, but he also suggested that implementing such a change required a supportive culture and good governance structures.

Based on these interviews, it is clear that the experts view *Lean* as part of a systemic approach to quality and service improvement, suggesting that it is more than implementing a 'tool-kit'. All experts focus on the managerial leadership role in owning *Lean*, coupled with achieving staff buy-in and training to achieve it in the widest sense of delivering internal organisational processes to eliminate waste and patient care as well as delivering externally focused objectives such as delivering patient satisfaction. All interviewees refer to the fact that initially external consultants were retained to implement *Lean* projects in Irish hospitals, but that Irish experts are now trained in *Lean healthcare* implementation.

Discussion

As expected and evidenced by our findings, *Lean* implementation in the Irish healthcare service can be considered to be piecemeal and sporadic. Some *Lean* processes and methodologies are being implemented, but there is no evidence of a systemic approach to *Lean* implementation across the sample of Irish hospitals.

Pettersen (2009), building on the work of Hines et al. (2004) and Shah and Ward (2007), developed a framework identifying approaches to *Lean* which provides a way of mapping *Lean* implementation in organisations. He distinguishes between: a) approaches towards *Lean* implementation, classified as performative (practical) or ostensive (philosophical); and b) level of *Lean* implementation, which he describes as discrete (operational) or continuous (strategic). This provides four distinctive categories of approaches to *Lean* implementation: 'toolbox *Lean*'; '*Leanness*'; 'becoming *Lean*'; and '*Lean* thinking'. Burgess (2012) utilised this framework to categorise her findings on *Lean* implementation in healthcare in the UK. We have adapted this framework slightly, reverting to the language originally suggested by Hines et al. (2004) regarding the level of operational implementation as operational and strategic, and that suggested by Shah and Ward (2007) regarding the approach to *Lean* as being philosophical or practical.

As can be seen in Table 6.1, the quadrants illustrate different potential approaches to *Lean* implementation. For example, a hospital adopting an operational, practical approach will, according to Burgess (2012), be involved in a set of specific projects including the Productive Ward, and will be using a tool-kit approach.

Applying this framework, our sample of Irish hospitals is predominantly categorised as adopting an operational, practical approach, described as a 'Toolbox *Lean*' approach. One Irish hospital adopts a philosophical approach to *Lean* implementation and can be assigned to the '*Leanness*' category, as it currently has a programme approach with a strategic objective to achieving a systemic approach. This approach is based on the vision of managerial leadership committing to integrating

Table 6.1 Adapted version of Pettersen's (2009) lean implementation framework

	Operational	Strategic
Philosophical	Leanness programme approach	Lean thinking systemic approach
Pratical	Tool-box lean projects PW	Becoming lean

Lean healthcare into the culture, structures and processes of the hospital, the appointment of a *Lean* specialist, and the establishment of a *Lean Academy* to communicate the vision and provide training in *Lean* processes.

This snapshot of the current *Lean* implementation situation in Irish hospitals appears to be consistent with the disjointed and fragmented approach to *Lean* found in healthcare organisations around the world and well documented in the literature (Poksinska 2010). There are of course exceptions, with the Virginia Mason Medical Center in the USA and Flinders Medical Centre in Australia being cited as examples of systemic *Lean* implementation in a healthcare context (see for instance: Bohmer and Ferlins 2006; Ben-Tovim et al. 2007). The main explanation for this approach, proffered in Ireland by the experts as well as in the literature, is that *Lean* as an approach to service and quality improvement is a relatively new phenomenon in the healthcare sector. Hines et al. (2004) suggest that we could consider health organisations as on a *journey*, evolving through stages of *Lean* development as set out in Table 6.1, from practical and operational to philosophical and strategic. Pettersen (2009) makes the insightful point that an internally focussed tool-kit approach favoured by practitioners, facilitating the development of 'pockets of best practice' (Radnor and Walley 2008) and described as 'islands of improvement' by our interviewees, should not be dismissed nor considered incorrect as these piecemeal interventions do achieve specific goals and have an impact. Further, in demonstrating small impacts (e.g. indirect savings within hospitals as outlined by our interviewees, and overcoming functional silos, viewed negatively by certain authors, e.g. Towill and Christopher 2005; Waldman and Schargel 2006), as evidenced in one hospital in our sample, these piecemeal approaches may in fact be developmental steps from a practical, operational tool-kit approach towards a philosophical, strategic systemic approach to service quality improvement. Other explanations put forward to explain the slow take-up and spread of *Lean* in Irish hospitals include the fact that the financial benefits of *Lean* for Irish hospitals appear to be mainly indirect, and the short tenure of senior management in Irish hospitals (White et al. 2014) results in a focus on short-term hits as opposed to long-term strategic change interventions.

Nevertheless, *Lean* theory and our interviewees advocate a more coordinated and systemic approach to *Lean* implementation (Burgess 2012). All interviewees recognise the relevance of context to achieve this. At the hospital level, the impact of financial constraints and their impact on the strategic choices managers can make, within the constraints of broader government policy, were mentioned. It was suggested by the experts that the Irish healthcare service, in seeking inspiration from best practice abroad, may lead *Lean* consultants and experts to overlook important contextual aspects crucial to the successful implementation of quality improvement in an Irish context. The interviewees in this study warn against a narrow, best-practice approach to service improvement and recommend the development of a structured implementation methodology tailored to the specific hospital, a view echoed by Stanton et al. (2014). Managerial leadership of the *Lean* process is widely acknowledged as imperative to successful implementation. Both literature (Al-Balushi et al. 2014) and interviewees underscore the importance of an appropriate organisational structure and culture, with the interviewees stressing the importance of empowering staff and securing staff buy-in into the process of change.

Conclusions and Limitations

Lean, as a recent strategic philosophical approach to service and quality improvement in healthcare organisations, offers the promise to streamline service provision from a patient-centred perspective and reduce waste across the health delivery system. The promise of these improvements, coupled with the strong prescriptive recommendation from both theory and practice to adopt a systemic approach, are recognised.

This snapshot of sporadic, piecemeal *Lean* implementation in a small sample of Irish acute hospitals has demonstrated that the pattern of *Lean* implementation in Ireland is similar to that reported in other countries. The Irish approach is described as practical and operational, evidencing some specific *Lean* projects and Productive Ward initiatives in Irish hospitals. Based on both the hospital annual report and the interviews with experts, there was evidence of strategic intent towards integrating a *Lean* philosophy into the service improvement processes in one hospital. We argue that these findings demonstrate that Irish hospitals are at the beginning of a *Lean* journey, and that with the leadership, training, supportive organisational structures and culture prescribed by *Lean* theorists and recommended by practitioners, this philosophical approach will develop. Then, the positive benefits to be accrued from this process innovation will be evident in better patient-centred service delivery and tangible cost savings.

Our study investigating the implementation of *Lean* in Irish acute hospitals has a number of limitations. The fact that annual reports from a relatively small sample of hospitals were analysed may be viewed as a limitation. However, when the recently created Hospital Groups are better established and integrated, a more representative sample of Irish hospitals could be surveyed. Second, recognising that the implementation of *Lean* is a journey, conducting longitudinal research or carrying out the analysis at two points in time, similar to the study by Burgess and Radnor (2013), would enable the progression of *Lean* implementation within the broader context of service improvement in the Irish health service to be estimated. Third, it is possible that the annual reports analysed in Phase 1 could be incomplete, distorted and/or biased. Hospitals might be using *Lean* tools and/or methodologies, but these might not be mentioned in their annual reports. Interviewing Hospital Group managers could address the issue of hospitals not compiling annual reports as encountered during the research process. Finally, it is recognised that the interview target group of three experts is a limitation, but at the time of the study, there was widespread recognition that the three experts were the main champions of *Lean* in the Irish healthcare system.

References

Al-Balushi, S., Sohal, A. S., Singh, P. J., Al Hajri, A., Al Farsi, Y. M., & Al Abri, R. (2014). Readiness factors for lean implementation in healthcare settings—A literature review. *Journal of Health Organization and Management, 28*(2), 135–153.

Ben-Tovim, D. I., Bassham, J. E., Bolch, D., Martin, M. A., Dougherty, M., & Szwarcbord, M. (2007). Lean thinking across a hospital: Redesigning care at the flinders medical centre. *Australian Health Review, 31*(1), 10–15.

Bohmer, R., & Ferlins, E. M. (2006). *Virginia Mason Medical Center,* Harvard Business School Case 606-044. Boston, MA: Harvard Business School.

Bossert, J. (2003). Lean and six sigma—Synergy made in heaven. *Quality Progress, 36*(7), 31–32.

Brandao de Souza, L. (2009). Trends and approaches in lean healthcare. *Leadership in Health Services, 22*(2), 121–139.

Brandao de Souza, L., & Pidd, M. (2011). Exploring the barriers to lean health care implementation. *Public Money & Management, 31*(1), 59–66.

Burgess, N. (2012). *Evaluating lean in healthcare.* Ph.D. thesis, University of Warwick.

Burgess, N., & Radnor, Z. (2013). Evaluating lean in healthcare. *International Journal of Health Care Quality Assurance, 26*(3), 220–235.

D'Andreamatteo, A., Ianni, L., Lega, F., & Sargiacomo, M. (2015). Lean in healthcare: A comprehensive review. *Health Policy, 119*(9), 1–13.

Graban, M. (2008). *Lean hospitals: Improving quality, patient safety, and employee satisfaction.* Boca Raton, FL: Productivity Press.

Grbich, C. (2007). *Qualitative data analysis* (1st ed.). Thousand Oaks, CA: Sage.

Health Service Executive. (2015). *Hospital groups.* Dublin: HSE. Retrieved from: http://www.hse.ie/eng/about/Who/acute/hospitalgroups.html. Accessed 12 Mar 2016.

Hines, P., Holweg, M., & Rich, N. (2004). Learning to evolve: A review of contemporary lean thinking. *International Journal of Operations & Production Management, 24*(10), 994–1011.

Mazzocato, P., Savage, C., Brommels, M., Aronsson, H., & Thor, J. (2010). Lean thinking in healthcare: A realist review of the literature. *Quality and Safety in Health Care, 19*(5), 376–382.

Pettersen, J. (2009). Defining lean production: Some conceptual and practical issues. *The TQM Journal, 21*(2), 127–142.

Pettigrew, A., & Whipp, R. (1991). *Managing change for competitive success.* Oxford: Blackwell.

Poksinska, B. (2010). The current state of lean implementation in health care: Literature review. *Quality Management in Healthcare, 19*(4), 319–329.

Proudlove, N., Moxham, C., & Boaden, R. (2008). Lessons for lean in healthcare from using six sigma in the NHS. *Public Money and Management, 28*(1), 27–34.

Radnor, Z., & Osborne, S. P. (2013). Lean: A failed theory for public services? *Public Management Review, 15*(2), 265–287.

Radnor, Z., & Walley, P. (2008). Learning to Walk Before We Try to Run: Adopting Lean for the Public Sector. *Public Money and Management, 28*(1), 13–20.

Shah, R., Chandrasekaran, A., & Linderman, K. (2008). In pursuit of implementation patterns: The context of lean and six sigma. *International Journal of Production Research, 46*(23), 6679–6699.

Shah, R., & Ward, P. T. (2007). Defining and developing measures of lean production. *Journal of Operations Management, 25*(4), 785–805.

Stanton, P., Gough, R., Ballardie, R., Bartram, T., Bamber, G. J., & Sohal, A. (2014). Implementing lean management/six sigma in hospitals: Beyond empowerment or work intensification? *The International Journal of Human Resource Management, 25*(21), 2926–2940.

Toussaint, J., & Gerard, R. (2010). *On the mend*. Cambridge, MA: Lean enterprise institute.

Towill, D. R., & Christopher, M. (2005). An evolutionary approach to the architecture of effective healthcare delivery systems. *Journal of Health Organization Management, 19*(2), 130–147.

Waldman, J. D., & Schargel, F. P. (2006). Twins in trouble (II): Systems thinking in healthcare and education. *Total Quality Management, 17*(1), 117–130.

Waring, J., & Bishop, S. (2010). Lean healthcare: Rhetoric, ritual and resistance. *Social Science and Medicine, 71*(7), 1332–1340.

White, M., Wells, J. S. G., & Butterworth, T. (2014). The impact of a large-scale quality improvement programme on work engagement: Preliminary results from a national cross-sectional-survey of the productive ward. *International Journal of Nursing Studies, 51*(12), 1634–1643.

Womack, J. P., & Jones, D. T. (1996). *Lean thinking: Banish waste and create wealth in your corporation* (1st ed.). New York: Simon and Schuster.

Part II

Embedding and Spreading Quality

7

Unlearning and Patient Safety

John G. Richmond

Introduction

Since the development of the patient safety movement in the early 2000s, healthcare organisations have moved forward with a plethora of safety improvement efforts. Whilst major advances have been made in the area of patient safety, it remains a significant and very real problem (Waring 2013). This affects patients in terms of unexpected injury, suffering and protracted care and healthcare organisations with regards to how to best configure services to deliver safer care. Unfortunately, despite current best efforts, it could be argued there is hardly any evidence of continued safety improvement (Landrigan et al. 2010).

Much has been invested in tools to promote organisational learning following incidents, such as Root Cause Analysis (RCA) (Nicolini et al. 2016). However, we know that hospitals rarely learn from their failures (Nicolini et al. 2011), and consequently, improvements based

J.G. Richmond (✉)
Warwick Business School, University of Warwick,
Coventry, UK
e-mail: Phd14jr@mail.wbs.ac.uk

© The Author(s) 2018
A.M. McDermott et al. (eds.), *Managing Improvement in Healthcare*,
Organizational Behaviour in Health Care,
https://doi.org/10.1007/978-3-319-62235-4_7

117

on learning from such failures are rarely implemented. This prevailir outcome is hereby referred to as an 'implementation gap'. Fresh wa of thinking are needed to improve upon past patient safety efforts address this gap in learning.

This chapter is about the importance of understanding, the unlear ing concept in the context of patient safety to ensure forward accour ability and the responsibility to learn lessons so that future people a not harmed by avoidable mistakes. This is particularly relevant to pr fessional organisations where new learning is often applied atop existir professional practices, establishing a need to first unlearn.

Unlearning, the discarding of obsolete organisational practices make room for new learning, is an under-researched concept and h been described by some health researchers as 'necessary to clear the wa for new (more appropriate) learning in healthcare practice' (Rushm and Davies 2004, p. 12). This chapter's updated unlearning model fil a research gap on the enactment of unlearning and considers the impo tance of cognitive, cultural and political factors that influence unlear ing in professional organisations.

The Patient Safety Agenda

The release of landmark government reports in both the USA an UK are largely responsible for developing the patient safety agenda i the Western world (Department of Health 2000; Kohn et al. 2000 The release of these reports led healthcare organisations to implemer patient safety initiatives. Unfortunately, there has been little evidenc of widespread safety improvement (Landrigan et al. 2010) as a result c this approach.

The health services literature is fairly comprehensive in documentin the trend of adverse events and medical errors in healthcare organisa tions across the globe, in this chapter, these are referred to collectivel as incidents. The proportion of inpatient visits leading to harmful inci dents ranges from a reported 3.7% in USA (Brennan et al., 1991) t 16.6% in Australia (Wilson et al. 1995) and as many as 70% of thes incidents are deemed preventable (Leape 1994).

A Promise to Learn: The UK's Response

A case could be made for taking a deeper look at one country, the UK, and its National Health Service's (NHS) efforts at attempting to learn from incidents. The NHS is an exemplary case given recent public calls for improved safety, resulting from several high-profile failures in care that resulted in government-led enquiries and calls for improvement.

The gap in learning from incidents remains an ever-present concern for both the public and government. As claimed by the UK Health Secretary(2015) Jeremy Hunt (2015), the NHS records 800 avoidable deaths every month, and 'wrong site surgery' incidents occur twice a week on average.

The UK's most recent efforts to bridge this gap in patient safety, to 'continually and forever reduce patient harm' (National Advisory Group on the Safety of Patients in England 2013, p. 5), come in the form of recommendations that propose transforming the NHS into a learning organisation by embracing an ethic of learning. Learning organisations ideally contain the following five characteristics: systems thinking, personal mastery, mental models, shared vision and team learning (Senge 1990). The NHS's vision is supported by the UK Health Secretary (Hunt 2015) who stated: 'The world's fifth largest organisation needs to become the world's largest learning organisation'.

The Implementation Gap

Numerous researchers have set out to analyse the initiatives undertaken by healthcare organisations to learn from incidents and prevent recurrences (Bishop and Waring 2011; Currie and Waring 2011; Iedema et al. 2008; Iedema et al. 2005; Nicolini et al. 2011; Wu et al. 2008; Vincent 2003). These studies have tended to emphasise the way in which incidents were analysed using tools like Root Cause Analysis (RCA), identification of risks and how lessons learned were shared using formal reports of recommendations for improvement.

A study which comprehensively focused on the use of RCA in practice has suggested that healthcare organisations rarely learn from their failures (Nicolini et al. 2011). This inability to learn has been hypothesised to be the result of several barriers, including a normalisation of deviance among staff (Vaughan 1999; Waring 2005), the promotion of quick fixes and workarounds rather than systematic analysis (Tucker and Edmondson 2003; Waring et al. 2007), and a predominant culture of blame (Carroll et al. 2002; Currie and Waring 2011; Department of Health 2000).

Figure 7.1 below is the learning circle used by the UK's Department of Health (2000) to conceptualise the process of organisational learning in response to incidents. It is shown here as a framework. Current approaches tend to reflect a 'find and fix' mindset (Hollnagel 2013, p. 6) resulting in a focus on the process of investigating incidents and compliance whilst skirting the issue of post-investigation learning and practice change.

Furthermore, new learning is often overlaid atop existing professional practices, making change difficult to embed and sustain, and highlighting the need to enact unlearning to make space for new safer practices.

Due to the impact of change initiatives on organisational matters such as resource allocation, authority and control, professional groups

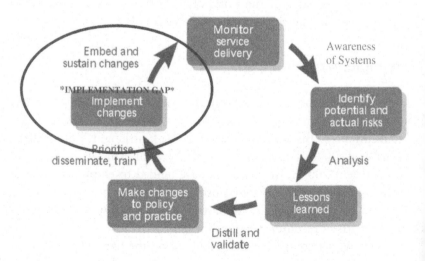

Fig. 7.1 Learning circle. Adapted from Department of Health (2000)

may be hesitant to unlearn past practices, and adopt new ones which threaten their organisational position. Freidson (1994) described this as collective control over knowledge traditionally associated with professional power and autonomy.

Enacting Unlearning

The concept of unlearning and how it might be enacted yields promise as a means to bridge the implementation gap by discarding obsolete practices. This chapter proposes unlearning as a concept worth critically exploring to understand how organisations can make room for new learning, which in the case of patient safety can result in improved, safer care.

Research grounded in unlearning literature has been limited in the healthcare setting. A ProQuest search of 36 separate databases for scholarly journals using the search terms 'unlearning' and 'healthcare' yielded 87 results, while 'unlearning' and 'patient safety' yielded only 8 results. No studies to date were found utilising unlearning to investigate patient safety.

The applicability of unlearning to the study and practice of patient safety is supported by Rushmer and Davies (2004) who highlight that 'getting people to *stop doing things* as well as getting new practices started' (p. 10, emphasis added) is a major challenge to managing quality, patient safety and medical error. This challenge results from clinician knowledge becoming stuck, ritualised and never removed from the organisation leading to the development of status quo (Rushmer and Davies 2004).

In contrast to research on learning, unlearning studies are scarce, resulting in a lack of knowledge about processes related to the concept, such as what forms it can take, how it occurs, and how it can be encouraged (Akgun et al. 2007; Becker 2005; Brook et al. 2015; Tsang and Zahra 2008).

The concept of unlearning first emerged in Hedberg's 1981 chapter on *How Organizations Learn and Unlearn* in the *Handbook of Organizational Design* (Nystrom and Starbuck 1981) where he wrote:

'knowledge grows, and simultaneously it becomes obsolete as reality changes. Understanding involves both learning new knowledge and discarding obsolete and misleading knowledge' (p. 3).

This chapter draws on Scott's (2008) institutional pillars in developing an updated model of unlearning. Given that this model centres around professionals, Scott's (2008) view of professionals as institutional agents, whose function 'can be described as variously specializing in creating, testing, conveying, and applying cultural-cognitive, normative, and/or regulative frameworks that govern one or another social sphere' (Scott 2008, p. 233), is applicable.

To develop this updated model, existing conceptualisations of unlearning are reviewed: fading, wiping and deep unlearning (Rushmer and Davies 2004), transformational unlearning (MacDonald 2002) and critical unlearning (Brook et al. 2015; Chokr 2009). Each of these conceptualisations of unlearning is explored within one of the updated model's three proposed dimensions—cognitive, cultural and political—drawn from Scott's (2008) cultural-cognitive, normative and regulative pillars.

Cognitive

Rushmer and Davies (2004) conceptualise unlearning as a cognitive process that occurs at three distinct levels: fading, wiping and deep unlearning. Whilst also viewing unlearning cognitively, MacDonald (2002) defines unlearning uniformly, as a transformative process that is complex, challenging and lengthy.

The idea of past learning automatically fading away or being forgotten is not as relevant to the updated model as other levels of unlearning which are deliberately and intentionally enacted. Similar to how safety recommendations are deliberately launched and do not occur automatically or without directed efforts, unlearning past practices won't happen by default. Wiping, as suggested by Rushmer and Davies (2004), is '[t]o be pushed into unlearning ... to be subject to focused, directive instruction to stop doing certain things' (p. 11), and '[t]o unlearn complex learning we might, therefore, need to be pushed or pulled down the unlearning curve' (p. 11). Moving along a learning curve, whilst useful conceptually, is a very cognitive activity, which makes it difficult to see and study ways to support a need for alternate perspectives, such as a practice-based approach.

Deep unlearning seems to only differ from other unlearning levels in the very rapid speed at which the unlearning curve is traversed (Rushmer and Davies 2004). This level could be seen as redundant in that it is also a deliberate enactment of unlearning, and its relevance exists only in proportion to the severity of the act necessitating unlearning.

Transformative unlearning (MacDonald 2002) is a more holistic conceptualisation, in that it considers the abandonment of established practices, knowledge and assumptions that may be linked to a sense of identity. In the case of MacDonald (2002), her identity as a nurse was challenged with the introduction of updated teaching guidelines pertaining to newborn supine, or side-lying positions. Transformative unlearning is a cognitive process of discernment involving being *receptive* to new evidence (despite fear of possible infant choking risks), *recognition* of evidence in support of new practices and *grieving* for the loss of identity attached to past practices (MacDonald 2002).

By moving past a cognition-oriented perspective, and incorporating practice-based elements that view unlearning as something which is enacted, an updated model of unlearning can overcome the limits of past models (Akgun et al. 2007; Rushmer and Davies 2004). A practice approach accepts the practices of organisational actors as a unit of analysis for understanding how learning and unlearning can occur (Nicolini 2013).

Questions remain about how organisational actors, such as professionals, discard practices. Tsang and Zahra (2008) provide no clear structure to define this discarding process. As a starting point, what factors influence the discarding of professional practices? What role might cultural and political factors play in unlearning professional practices?

Cultural

Whilst settling on a definition of culture can be difficult, one review found 12 different definitions and was able to highlight two theoretical features common to most—the use of the word 'shared', and a reference to culture as unique to a particular context (Martin 2002). To understand the relationship between culture and unlearning, the case of Bristol Royal Infirmary (BRI) is reviewed. This provides an example where culture enforced questionable professional practices that inhibited new learning (Weick and Sutcliffe 2003).

The BRI paediatric cardiac surgery programme tragically had much higher mortality rates (32.2%) than other similar hospitals in the UK (21.2%) (Weick and Sutcliffe 2003). These problems were said to stem from a 'culture of entrapment', which is 'the process by which people get locked into lines of action, subsequently justify those lines of action, and search for confirmation that they are doing what they should be doing' (Weick and Sutcliffe 2003, p. 73). The culture at BRI trapped professionals into behavioural commitments which saw them, justify and rationalise poor performance stemming from a supposedly high volume of unusually complex patient cases, rather than considering their own failings or systematic issues (Weick and Sutcliffe 2003).

That culture led to an ossification of professional practices related to justification and rationalisation is evident in the case of BRI, highlighting the importance of unlearning. Overcoming this would have required unlearning practices associated with the prevailing mindset: 'it would have taken a different mindset ... It would have required *abandoning* the principles which then prevailed' (Department of Health 2002, p. 4, emphasis added).

The relationship between culture and unlearning, in this case, seems to suggest that certain types of culture (i.e. a culture of entrapment) reinforce a prevailing mindset which prevents professionals from unlearning practices. For example, it was common practice following an incident for BRI surgeons to rely upon their own operation logs as the most reliable source of data for finding plausible justification, rather than also considering the interdependencies and perspectives of other hospital staff (Weick and Sutcliffe 2003).

In the case of BRI, a culture of entrapment played a role in preventing deliberate unlearning from being enacted and is therefore suggestive of a negative relationship between the concepts. This raises questions concerning what type of culture might support the enactment of unlearning.

We know that the implementation of Root Cause Analysis (RCA) practices in organisations can lead to changes in culture, which result in more trust and openness among staff, nurture more disciplined thinking about problems in the organisation (Carroll et al. 2002) and facilitate a more open safety culture that actively seeks out previous experiences of error in an effort to ensure they do not happen again (Department of Health 2001; Leape et al. 1998). Whilst a safety culture seems compatible with the enactment of unlearning, given the lack of research in this area it is difficult to say for certain, supporting the need for future studies that include a more robust model of unlearning incorporating culture.

Political

A weakness of the unlearning literature is a lack of emphasis on possible political factors which can influence unlearning. The importance of political influences on learning is brought to the fore by Contu and Willmot (2003), who explore a situated understanding of learning, which implicates learning in broader social structures involving relations of power. This updated model aims to incorporate these elements, to demonstrate how 'learning processes are inextricably implicated in the social reproduction of wider institutional structures' (Contu and Willmot 2003, p. 294).

To critically examine the unlearning concept it must be viewed as part of a wider learning literature that includes considerations of a social and political nature. This 'learning discourse' is the meaningful and structured totality of the subject of learning where organisational learning connects learning to organisation and has implications for the link between the wider social arena and organisations within which learning occurs (Contu et al. 2003).

In this context, learning is seen as an inevitable response to the uncertain and changing times of a globalised knowledge-based economy. This response is based on the premise that learning is uncritically recognised as a good thing, where any concept bearing a title which includes 'learning' is seen positively, such as a 'learning organization' (Contu et al. 2003, p. 933). What this emphasises is the dominance of 'learning' and its power as a tool in a wide range of social and political settings, as demonstrated by the UK's endorsement of becoming a 'learning organisation' as the solution to their NHS's safety woes.

Certain professionals such as doctors may view learning initiatives negatively and be hesitant to accept new learning, since they are bombarded with information regularly and experience reform fatigue. This results in new learning adding to rather than replacing old practices. Whilst predominantly viewed as positive, learning conceals constraints on what can be learned, both socially and organisationally, which are both controlling and controlled (Contu et al. 2003).

By considering the political influences that may weigh on the enactment of unlearning, we bring a critical perspective to the updated model. As some researchers have suggested, unlearning is a way to enable a critical and unlearning attitude, where broader ideologies and practices are challenged (Brook et al. 2015; Chokr 2009). By adopting a critical attitude, organisational actors can differentiate between individual experience and political factors which influence the organisational challenges they face.

Critical unlearning, in contrast to inward focused deep and transformative unlearning, is an outward focused, liberating process. It involves critical reflection at both a collective and public level, and enables the questioning of dominant ways of thinking and rediscovery of subjugated knowledge (Brook et al. 2015; Chokr 2009). A key characteristic of critical unlearning is its social focus, not on the motivations and actions of individuals, but on organisational and institutional forces which impact upon the situation. Thus, it frames the processes of working, managing and learning in organisations in a context of wider social influences.

Critical unlearning is a means to challenge the 'relentlessly performative' nature of learning by questioning underlying dominant knowledges and social ideals. This questioning attitude empowers organisational actors with a 'desire and willful determination *not to be taken in*' (Chokr 2009, p. 6), leading to the rediscovery of previously suppressed knowledges outside the governing variables of the organisation.

For example, the process of learning from medical errors can be constrained by Root Cause Analysis (RCA) (Peerally et al. 2016). RCA is prone to political hijacking, which stems from investigative processes that lack independence from the organisation where the error took place, amongst other factors. There is also a risk of investigative reports in themselves, rather than learning and improvement, becoming a goal of RCA. Furthermore, RCA reports can end up tailored to moderate partisan interests, hierarchical tensions and interpersonal relationships (Peerally et al. 2016). Thus, cultivating a critical attitude towards RCA can empower organisational actors to consider these extraneous shortcomings, and begin a journey towards effective organisational learning.

Research Agenda

Whilst this chapter presents the idea of unlearning as holding value for researching and managing patient safety, the literature suffers from a lack of enquiry beyond initial descriptions, and no focused attempts, with the exception of Brook et al. (2015), to place the process of unlearning in the broader literature on learning and organisations.

Conceptualising Unlearning

Based on the dimensions of unlearning reviewed above, an updated conceptual model has been constructed (see Fig. 7.2). This model provides a framework for researchers to carry out further research on how unlearning is enacted in professional organisations.

The updated unlearning model highlights the cognitive, cultural and political dimensions across which unlearning might be enacted by organisational actors, at an individual and collective level. The factors which are implicit to the process of unlearning are identified for each dimension. Unlearning of the deliberate and transformative type is enacted at the individual and organisational levels, whilst critical unlearning of exogenous factors occur at the political and environmental levels.

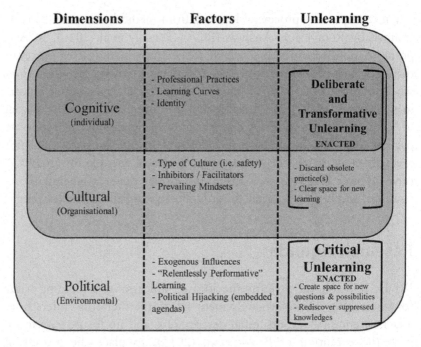

Dimensions	Factors	Unlearning
Cognitive (individual)	- Professional Practices - Learning Curves - Identity	**Deliberate and Transformative Unlearning** ENACTED
Cultural (Organisational)	- Type of Culture (i.e. safety) - Inhibitors / Facilitators - Prevailing Mindsets	- Discard obsolete practice(s) - Clear space for new learning
Political (Environmental)	- Exogenous Influences - "Relentlessly Performative" Learning - Political Hijacking (embedded agendas)	**Critical Unlearning** ENACTED - Create space for new questions & possibilities - Rediscover suppressed knowledges

Fig. 7.2   A practice-based framework for researching unlearning

A goal of further research should be to validate and explore this model's potential in a professionalised setting, like healthcare, to improve upon patient safety practice and research. The purpose of this section is to highlight what patient safety researchers may wish to consider in studying the concept, to advance theory in this area and translate knowledge to practitioners on the front lines of healthcare.

By moving from a cognitive perspective to a practice-based approach to unlearning, the updated model views the routinized practices of professionals as a unit of analysis for understanding how learning and unlearning can occur (Nicolini 2013). Since it pertains to observing unlearning, the discarding of practices, and assuming a general desire to understand how the phenomena occur, what enables and inhibits it, a starting point is examining the practices of professionals in organisations. Compatible with this approach is a desire to access professionals' 'logic of practice', to build theory which better reflects the way in which practices are enacted (Sandberg and Tsoukas 2011).

As suggested by Sandberg and Tsoukas (2011), examining temporary breakdowns, such as interruptions or disturbances in the flow of practice, emphasises a 'focus on … the sociomaterial practice (i.e. ourselves, others, and tools) as something separate and discrete, singling people and tools out from their relational whole' (p. 344). It is during these breakdowns that professionals' absorbed coping is disrupted and, momentarily, the entirety of the sociomateriality of practice, that is the entanglement of the social and material, is observable (Sandberg and Tsoukas 2011).

The healthcare setting, especially scenarios involving patient safety, offers many opportunities to observe practice breakdowns, in the form of professional response to medical errors, incidents, and Root Cause Analysis (RCA) investigations. Analysing breakdowns in professionals' practice offer researchers an opportunity to assemble ideas about how practices might be discarded. Drawing on work from the military field involving friendly fire (Snook 2000), it is possible to identify the 'practical drift' (Snook 2000, p. 225) that occurs during incidents. In Snook's (2000) analysis, this resulted when local practices drifted and no longer conformed to formal procedures.

Adopting a practice-based approach helps to ensure the updated model of unlearning reflects how 'organizational practices are constituted and enacted by actors' and 'capture essential aspects of the logic of practice' (Sandberg and Tsoukas 2011 p. 339). This approach will develop unlearning as a concept, making it more applicable to the practices of front-line healthcare professionals, thus helping researchers in this field bridge the gap between theory and practice.

Conclusion

This chapter adds to the scarce but growing body of literature on unlearning by contributing an updated model as a framework for how this concept can be enacted in the context of patient safety, and more broadly in professional organisations. The intent of this conceptual chapter has been to focus attention on advantages inherent in enacting unlearning for practitioners and researchers involved in patient safety.

The patient safety agenda was reviewed and the UK's 'learning organisation' solution for patient safety discussed. The implementation gap was identified and unlearning proposed as a solution to overcome this gap. Unlearning is a specific type of learning that is enacted to ensure obsolete professional practices are removed, creating space to embed new learning. The cognitive nature of the past unlearning literature was discussed and the need to adopt a practice-based approach for future research presented. The potential relationship between culture and politics on the enactment of unlearning were also reviewed and incorporated into an updated unlearning model for further study.

This chapter serves as a reminder for those involved in patient safety to consider the broader context in which their efforts are placed. As a research agenda, this chapter provides a starting point for thinking about how unlearning can be studied in organisations.

Acknowledgements The author of this chapter is funded by the NIHR CLAHRC West Midlands Initiative. This chapter presents independent research and the views expressed are those of the author and not necessarily those of the NHS, the NIHR or the Department of Health.

References

Akgun, A., Byrne, J., Lynn, G., & Keskin, H. (2007). Organisational unlearning as changes in beliefs and routines in organisations. *Journal of Organisational Change Management, 20*(6), 794–812.

Becker, K. (2005). Individual and organisational unlearning: Directions for future research. *International Journal of Organisational Behaviour, 9*(7), 659–670.

Bishop, S., & Waring, J. (2011). Exploring the contributions of Professional-practice networks to Knowledge sharing, problem-solving and patient safety. In E. Rowley & J. Waring (Eds.), *A socio-cultural perspective on patient safety* (pp. 171–188). Surrey: Ashgate Publishing Limited.

Brennan, T., Leape, L., Laird, N., Hebert, L., Localio, A., Lawthers, A., & Hiatt, H. (1991). Incidence of adverse events and negligence in hospitalized patients: Results of the Harvard Medical Practice Study I. *The New England Journal of Medicine, 324,* 370–376.

Brook, C., Pedler, M., Abbott, C., & Burgoyne, J. (2015). On stopping doing those things that are not getting us to where we want to be: Unlearning wicked problems and critical action learning. *Human Relations, 69*(2), 1–21.

Carroll, J. S., Rudolph, J. W., & Hatakenaka, S. (2002). Lessons learned from non-medical industries: Root cause analysis as culture change at a chemical plant. *Quality and Safety in Healthcare, 11,* 266–269.

Chokr, N. (2009). *Unlearning: Or how not to be governed?* Exeter: Societas Imprint Academic.

Contu, A., Grey, C., & Örtenblad, A. (2003). Against learning. *Human Relations, 56*(8), 931–952.

Contu, A., & Willmott, H. (2003). Re-Embedding situatedness: The importance of power relations in learning theory. *Organisation Science, 14*(3), 283–296.

Currie, G., & Waring, J. (2011). The politics of learning: The dilemma for patient safety. In E. Rowley & Waring, J. (Eds.), *A socio-cultural perspective on patient safety* (pp. 171–188). Surrey: Ashgate.

Department of Health. (2000). *An organisation with a memory: Report of an expert group on learning from adverse events in the NHS.* London: The Stationery Office.

Department of Health. (2001). *Building a safe NHS for patients.* London: The Stationery Office.

Department of Health. (2002). *Learning from Bristol.* Retrieved from https://www.gov.uk/government/uploads/system/uploads/attachment_data/file/273320/5363.pdf.

Freidson, E. (1994). *Professionalism reborn: Theory, prophesy and policy.* Cambridge: Polity Press.

Hedberg, B. (1981). How organisations learn and unlearn. In P. C. Nystrom & W. H. Starbuck (Eds.), *Handbook of organisational design* (Vol. 1, pp. 3–27). Oxford: Oxford University Press.

Hollnagel, E. (2013). Making health care resilient: From safety-I to safety-II. In E. Hollnagel, J. Braithwaite, & R. Wears (Eds.), *Resilient healthcare* (Vol. 1). Farnham, Surrey: Ashgate.

Hunt, J. (2015). *Making healthcare more human-centred and not system-centred.* London: King's Fund. Retrieved from https://www.gov.uk/government/speeches/making-healthcare-more-human-centred-and-not-system-centred. Accessed 05 Dec 2016.

Iedema, R., Jorm, C., & Braithwaite, J. (2008). Managing the scope and impact of root cause analysis. *Journal of Health Organisation and Management, 22*(7), 569–585.

Iedema, R., Jorm, C., Long, D., Braithwaite, J., Travaglia, J., & Westbrook, M. (2005). Turning the medical gaze in upon itself: Root cause analysis and the investigation of clinical error. *Social Science and Medicine, 62,* 1605–1615.

Kohn, L. T., Corrigan, J. M., & Donaldson, M. S. (Eds.). (2000). *To Err is human: Building a safer health system.* Washington, DC: National Academy Press.

Landrigan, C. P., Parry, G. J., Bones, C. B., Hackbarth, A. D., Goldmann, D. A., & Sharek, P. J. (2010). Temporal trends in the rates of patient harm resulting from medical care. *New England Journal of Medicine, 363*(2), 124–134.

Leape, L. (1994). The preventability of medical injury. In S. M. Bogner (Ed.), *Human error in medicine* (pp. 13–27). Hillsdale, NJ: Lawrence Erlbaum Associates.

Leape, L., Woods, D., Hatlie, M., Kizer, K., Schroeder, S., & Lundberg, G. (1998). Promoting patient safety by preventing medical error. *JAMA, 280*(16), 1444–1447.

MacDonald, G. (2002). Transformative unlearning: Safety, discernment and communities of learning. *Nursing Inquiry, 9*(2), 170–178.

Martin, J. (2002). Pieces of the puzzle: What is culture? What is not culture? In J. Martin, *Organisational culture: Mapping the terrain (Foundations for organizational science)* (pp. 55–93). Thousand Oaks, CA: Sage.

National Advisory Group on the Safety of Patients in England. (2013). *A promise to learn—A commitment to act.* Retrieved from https://www.gov.uk/government/uploads/system/uploads/attachment_data/file/226703/Berwick_Report.pdf. Accessed 05 Dec 2016.

Nicolini, D. (2013). *Practice theory, work, & organisation: An introduction.* Oxford: Oxford University Press.

Nicolini, D., Mengis, J., Meacheam, D., Waring, J., & Swan, J. (2016). Is there a ghost in the machine? Recovering the performative role of innovations in the global travel of practices. In J. Swan, S. Newell, & D. Nicolini (Eds.), *Mobilizing knowledge in healthcare*. Oxford: Oxford University Press.

Nicolini, D., Waring, J., & Mengis, J. (2011). Policy and practice in the use of root cause analysis to investigate clinical adverse events: Mind the gap. *Social Science and Medicine, 73*, 217–225.

Nystrom, P. C., & Starbuck, W. H. (Eds.). (1981). *Handbook of organisational design*. Oxford: OUP.

Peerally, F., Carr, S., Waring, J., & Dixon-Woods, M. (2016). The problem with root cause analysis. *BMJ Quality & Safety*. doi: 10.1136/bmjqs-2016-005511.

Rushmer, R. K., & Davies, H. T. O. (2004). Unlearning in health care. *Quality and Safety in Health Care, 13*(Suppl. 2), ii10–ii15.

Sandberg, J., & Tsoukas, H. (2011). Grasping the logic of practice: Theorizing through practical rationality. *Academy of Management Review, 36*(2), 338–360.

Scott, R. (2008). Lords of the dance: Professionals as institutional agents. *Organization Studies, 29*(2), 219–238.

Senge, P. (1990). *The fifth discipline: The art and practice of the learning organization*. London: Doubleday.

Snook, S. (2000). *Friendly fire*. Princeton, NJ: Princeton University Press.

Tsang, E. W. K., & Zahra, S. A. (2008). Organisational unlearning. *Human Relations, 6*(10), 1435–1462.

Tucker, A. C., & Edmondson, A. C. (2003). Why hospitals don't learn from failures: Organizational and psychological dynamics that inhibit system change. *California Management Review, 45*(2), 55–72.

Vaughan, D. (1999). The dark side of organisations: Mistake, misconduct, and disaster. *Annual Review of Sociology, 25*, 271.

Vincent, C. A. (2003). Understanding and responding to adverse events. *The New England Journal of Medicine, 348*, 11.

Waring, J. (2005). Beyond blame: Cultural barriers to medical incident reporting. *Social Science and Medicine, 60*, 1927–1935.

Waring, J., Harrison, S., & McDonald, R. (2007). A culture of safety or coping? Ritualistic behaviours in the operating theatres. *Journal of Health Services Research and Policy, 12*(Suppl. 1), 3–9.

Waring, J. (2013). What safety-II might learn from the socio-cultural critique of safety-I. In E. Hollnagel, J. Braithwaite, & R. Wears (Eds.), *Resilient healthcare* (Vol. 1). Farnham, Surrey: Ashgate.

Weick, K. E., & Sutcliffe, K. M. (2003). Hospitals as cultures of entrapment: A re-analysis of the Bristol Royal Infirmary. *California Management Review*, 45(2), 73–84.

Wilson, R., Runciman, W., Gibberd, R., Harrison, B., Newby, L., & Hamilton, J. (1995). The Quality in Australian healthcare study. *The Medical Journal of Australia*, 163, 458–471.

Wu, A. W., Lipshutz, A. K., & Pronovost, P. J. (2008). Effectiveness and efficiency of root cause analysis in medicine. *The Journal of the American Medical Association*, 299, 6.

8

Checklist as Hub: How Medical Checklists Connect Professional Routines

Marlot Kuiper

Introduction

In 1935, the aviation industry introduced the use of checklists to prevent human mistakes. That year the US Army Air Corps invited aeroplane manufacturers to build its next-generation long-range bomber. In theory, this 'competition' between two rivals, Boeing and Martin & Douglas, was expected to be a mere formality. Boeing was far ahead and its design had conquered any other design; the result of the competition seemed a foregone conclusion. However, during the test flight with a very experienced pilot, the innovative Boeing aircraft crashed and exploded. Two out of five crew members died. An investigation revealed that the crash had been due to pilot error. The innovative design required the pilot to perform several complex tasks, more than ever before. All in all, the new Boeing was deemed 'too much airplane

M. Kuiper (✉)
Utrecht School of Governance (USG), Utrecht University,
Utrecht, The Netherlands
e-mail: m.kuiper@uu.nl

© The Author(s) 2018
A.M. McDermott et al. (eds.), *Managing Improvement in Healthcare*,
Organizational Behaviour in Health Care,
https://doi.org/10.1007/978-3-319-62235-4_8

for one man to fly'. Martin & Douglas won the competition with their smaller, less advanced aircraft, and Boeing nearly went bankrupt.

However, the US Army Air Corps still decided to purchase a few Boeing planes, and they came up with a very simple design to deal with their complexity: they designed a pilot's checklist that included step-by-step checks before take-off and landing. With the checklist, pilots managed to perform 1.8 million flights without any accident. The Boeing turned out to be the craft that gained the US the greatest advantages in the air during the Second World War. The rest is history, and the checklist became routine practice within the aviation industry (based on Gawande 2010).

The successes with checklists in the pioneering aviation industry made other sectors adopt the concept of checklists too. In many cases, this was done successfully. For example, the chemical and engineering industries integrated checklists into their daily work processes (e.g. Braham et al. 2014; Thomassen et al. 2011). However, the medical field remarkably lagged behind in this development. Despite many serious and thorough attempts—for example, resulting in a checklist that lists crucial safety checks before surgery—the medical profession still reports compliance rates that do not exceed 'average'[1] (e.g. Rydenfält et al. 2013; Van Klei et al. 2012). Newspapers report that 'not all surgeons follow checklists that prevent bad mistakes', even though the simplicity of the checklist is often emphasized. What results is that medical doctors feel assaulted by reprimands like 'they just don't do it!' Explanations for unsatisfactory compliance rates in this domain often emphasize the characteristics of the medical professional culture, with professionals who are not very susceptible to change, and strongly rely on their institutionalized autonomy (e.g. Freidson 1994; Tunis et al. 1994). A lack

---

[1]Although it must be said that compliance rates in studies that use self-registration data are a lot higher, sometimes even up to 99 or 100% (see e.g. Urbach et al. 2014; Fourcade et al. 2011). However, observational studies report compliance rates that hover around 30 per cent (complete checklist compliance) to 55% (partial checklist compliance) (e.g. Rydenfält et al. 2013; Van Klei et al. 2012). In later paragraphs we will further reflect on consequences of these different study designs.

of motivation is often considered one of the most important barriers to implementation (e.g. Cabana et al. 1999). Although some of these claims indeed partly explain the poor use of checklists—the image of the medical profession as 'stubborn' and not open to change did not come out of nowhere—in this chapter it is claimed that there is more to this picture as one broadens its scope. I will do this by looking at professional routines. This chapter specifically focuses on how an envisioned routine—a safety checklist—*interacts* with existing routines by presenting the critical case of the Surgical Safety Checklist (SSC). There are two main reasons why this perspective is relevant for studying a checklist in this medical domain. First of all, surgical care delivery can be viewed as a complex web of multiple interdependent professional routines. Next, and adding to this, the SSC was explicitly designed to *connect* a number of these routines. Thus, in order to understand why a checklist becomes routine practice or not, we explicitly need to consider its relation with other routines. The research question central to this chapter is: '*How does a checklist interact with existing routines and how does this affect the creation of a connective routine?*'

Professional Routines

The study of organizational routines has boomed the past few years, especially since Feldman and Pentland (2003) first associated routines with organizational change. In classical work on organizational routines scholars predominantly associated them with organizational stability (e.g. Cyert and March 1963; Nelson and Winter 1982), and therefore, also with inertia and even mindless behaviour. Feldman and Pentland (2003) challenged this traditional view by conceptualizing internal routine dynamics and discerning two key routines dimensions: ostensive and performative.

The ostensive dimension is the abstract, generalized idea of the routine, used to refer to a certain activity or justify what people do. It relates to structure. The *performative* dimension consists of 'actual performances by specific people, at specific times, in specific places'.

It relates to agency. In other words, the ostensive dimension is the idea, the performative dimension is the enactment (Feldman and Pentland 2003). Third, the authors distinguish artefacts as factors that enable or constrain elements of routines. These artefacts take on visible and tangible forms, like protocols and checklists. Feldman and Pentland recognized a recursive cycle of performative and ostensive aspects, also affected by artefacts. The dynamic of the two produces both stability and change.

Since the recognition of internal routines dynamics, scholars have attempted to unravel internal routine dynamics to analyse how routines evolve over time. Though the basic idea that routines occur in 'bundles' has been recognized for many years (e.g. Nelson and Winter 1982). This idea indicates the need to consider the *multiplicity* of routines. However, 'we have studied stability and change in individual routines, but there has been less focus on how routines *affect one another* and how they work together to support stability and change' (Feldman et al. 2016, p. 509, emphasis added).

Moreover, very little is known about the interaction of routines in high-complexity *professional* domains. In this chapter I aim to fill this gap, by explicitly studying how a checklist—thus an envisioned connective routine—interacts with existing routines and how this affects the emergence of such a connective routine. Most studies conducted on checklist use in medical domains analysed the specific routine of a checklist in isolation from its context among other routines. For example, studies only report numbers on the self-registration of checklist use, and the few observational studies that have been conducted merely observed the performance of the specific checklist without taking other routines into consideration (e.g. Rydenfält et al. 2013; Pickering et al. 2013; Levy et al. 2012). In this way, we only get to see if a specific checklist has been used, but not how other routines affected its performance.

In this chapter, the framework of Feldman and Pentland is extended as presented in Fig. 8.1. The assumption is that the routine that emerges in the middle—the envisioned connective routine—is formed by

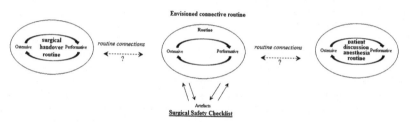

Fig. 8.1 Surgical safety checklist as 'hub' connecting multiple professional routines

its interaction with other routines. A focus on the *interaction* of routines is especially relevant for the study of routines in complex professional healthcare settings, since care delivery consists of a multiplicity of interdependent professional routines (e.g. patient handovers, anaesthetic routines) that need to come together in the multidisciplinary team checklist routine. Put differently, surgical care is not only about coordinating a series of related routines within a sub-discipline, it is also about ongoing coordination with professional routines that shape the work in other sub-disciplines such as anaesthesia. During the surgical routine, the surgeon draws on professional knowledge to continuously assess what has been done and what still needs to be performed, which involves ongoing coordination with other routines such as those in anaesthesia. The performance of such professional routines is thus highly interdependent and entails coordinating a series of connections with related routines (Hilligoss and Cohen 2011).

However, the creation of such connective routines might be difficult, for at least two reasons. First, the artefact explicitly prescribes behaviour, while established professional routines are mostly implicit—encompassing tacit knowledge. Although these routines structure work, they are not backed up by codified artefacts. Second, professional routines are mostly segmented. Socialization into sub-disciplines makes professionals construct a sense of *their* profession which includes its duties, boundaries, values, aspirations and relation to others (Abbott 1988; Freidson 1994; Cruess et al. 2015). Different routines, therefore, guide behaviour in the various sub-disciplines.

Medical Checklists

A checklist is most commonly claimed to be a 'memory aid', and consequently, a 'simple intervention'. As reflected by one of the introductory newspaper headings, a surgical safety checklist is sometimes even seen as an individual tool, for a surgeon who has to comply with a rule. Although in the scholarly literature checklists are indeed seen as tools for surgical teams, scholars in the field of implementation science often approach checklists as a technical intervention, not acknowledging the context in which they have to be applied. For example, in their review in the British Medical Journal Quality and Safety, Treadwell et al. (2013: 1) conclude that 'surgical checklists represent a relatively simple and promising strategy'. The way in which (safety) checklist are approached in these studies echoes a rather strong form of technical determinism (Pentland and Feldman 2008) 'Designing' new routines would be a simple matter of creating the checklist, and once in place, these written checklists will determine patterns of action: they will get checked. This relates to the claim made by Atul Gawande (2010), who stated: 'The checklist works—as long as it is implemented well.'

However, checklists are far from simple procedural tools. They are *social* interventions that interfere with both the practical and social taken for granted ways of working (see also Bosk et al. 2009). From a more sociological perspective, scholars have investigated why professionals tend to *resist* checklist that prescribe action patterns (e.g. Evetts 2002; Bosk et al. 2009). However, studies have shown that professionals not only work against reforms but also *with* reforms (e.g. Wallenburg et al. 2016). There is a call for more nuance than organizational control or professional resistance (ibid.; Waring and Bishop 2013) In this chapter I aim to provide such a nuanced perspective by tracing at a micro level how routines are created or changed through everyday mundane practices.

The case central to this chapter, the SSC, was explicitly designed to create connections between different professional routines, or as one of the respondents stated: 'everything has to come together'. The SSC consists of three parts: a morning *briefing* at 8 a.m. in which all patients are discussed by the whole surgical team; and two patient-related moments: a *time out* right before incision; and a *sign out* just before the patient leaves the operating theatre.[2] For example, in the time out the complete surgical team has to perform the latest safety check, in which they rely on each other's information. The professional routines of the sub-disciplines thus have to connect in this checklist routine. I will empirically explore how the various professional routines connect (or not) in the checklist routine, and therefore take into consideration other firmly established routines.

Empirical Research

The research aim was to get a contextualized understanding of checklists in professional domains by studying how various professional routines interact. Therefore, I adopted a focused ethnographic (FE) approach (see e.g. Higgingbottom et al. 2013). 'Focused' in this approach refers to a problem orientation; within FE the topics of enquiry are pre-selected. Although the focus of this study was clearly demarcated in advance—the Surgical Safety Checklist—this qualitative method, using an inductive paradigm to gain in-depth understanding, differs from deductive (observational) studies that might fail to capture a holistic perspective. This FE approach allowed for studying how a checklist is embedded within daily work routines.

The author was appointed as a 'research assistant' in the hospital under study, and with this employment formal access to the field was arranged. Focused ethnography is characterized by episodic observation. Because of its problem orientation, the purpose is not to 'go native' but

---

[2]The World Health Organization introduced the first version of this Surgical Safety Checklist, and explicitly encouraged hospitals to adapt this general format to their local circumstances. Therefore, the hospital under study transformed the 'sign in' check to a morning 'briefing' in which all patients of the day are discussed. More information on the Surgical Safety Checklist can be found on the WHO website.

to get an in-depth understanding of the selected study topic. In the course of 8 months, approximately five full working weeks were spent in the surgery department for observation. These observations were preceded by informal interviews about the research topic with all the respondents who consented to observations.

Since the aim was to find out how the SSC connects to other routines, I did not merely observe the performance of the checklist in the operating theatre, as most research so far has done. In addition, I observed the full working days of different professionals who were involved in the checklist routine to get to know the various routines they were engaged in, as well as the interaction of these routines. I used a shadowing technique to do this (McDonald 2005). I shadowed both specialist surgeons and anaesthesiologists to learn about routines from different professional perspectives.

During observation, detailed field notes were taken. Data collection was extended by recording summaries of many informal conversations and obtaining various related artefacts such as policy documents, checklists, emails and information from the software system. Data analysis consisted of thematic analysis of the detailed written field notes and conversation reports using NVivo software.

The ethnographic field notes taken during observation were jotted down in a notebook and meticulously written up in digital format after every episode of data collection. Both observation and conversation data were imported into NVivo10 software for the purpose of thematic content analysis. The analysis was based on an initial coding scheme developed from the conceptual model (Fig. 8.1), incorporating emergent themes as they were identified throughout the research process. During the coding process, themes were identified to describe both the actions and abstract ideas of the various team members and the circumstances under which connections emerged. They were used to explore the processes of connective routines as social, situated and ongoing activities.

Checklists in Action

Varying Checklist Performances

During episodic observations at the surgery department, I got to see many performances of the Surgical Safety Checklist.[3] A first finding was that from all these attended checklist performances, not one repetition of the checklist routine was the same. The routine performances strongly varied, for example, in the number of people that attended, how fluently the routine fit within the process, how extensively the checklist was discussed, the extent to which participants paid attention, and who led the conversation. In other words, the connections as envisioned by the checklist were not always self-evidently established.

By shadowing different clinicians from different specialisms, I got to know the various routines they engaged in. As I learned about the interaction of routines, clues about the varying checklist-routine performances became evident. Based on the observation data, I first schematized an ideal typical situation in which the checklist does generate connections between different routines (Fig. 8.2). Although this visualization is a significant simplification of reality, it does provide insight into both the various practices that construct professional work and the envisioned connections.

The vertical flow of boxes represents the various activities individuals are engaged in. The horizontal lines in the figure represent the location in the processes where the different phases of the checklist (briefing, time out, sign out) have to be performed, and thus connections established.

There are a few important observations supplementing this visual. First, professional work is *layered* since it consists of: (1) individual work practices such as checking upon patients, (2) professional

---

[3]Depending on the perspective of observation—shadowing either a surgeon or an anaesthesiologist—the number of attended performances of the checklist in a day varied from five, in the case of a surgeon who had to perform two complex vascular surgeries (one briefing, two time outs and two sign outs), to 24, when shadowing an anesthesiologists who had to take care of anaesthesia for seven operations in OR1 and four in OR2 (two briefings, eleven time outs, eleven sign outs).

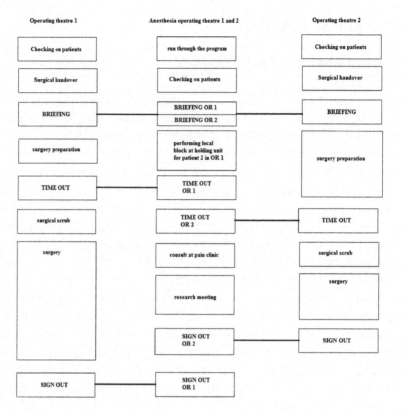

Fig. 8.2 Envisioned routine connections

routines within sub-disciplines, such as handovers; and (3) multidisciplinary routines that connect the various routines, such as the time out in the SCC. Second, the *organization* of work processes differs among the professional disciplines: the organization of surgical care is linear, whereas the organization of anaesthesia is entwined. Anaesthesiologists have to manage at least two linear surgical processes in different ORs simultaneously.

This figure merely represents one series of routines—one surgery in each operating theatre—while the number of operations per theatre can add up to seven or eight a day. Also, the blocks that represent time slots are clearly demarcated, but in reality the length of these blocks is highly unpredictable. The scheduled time for a surgery might be one hour, but because of unexpected events, for example concerning the patient's condition, this timing might fluctuate. Finally, this visual does not provide any information about the ostensive dimension of the various routines, and thus the values and norms encompassing these routines. It, therefore, neglects value judgements and thus pressures for prioritization.

All in all, the lines that represent the connections in the ideal type are not that straightforward. In reality, the envisioned connections lead to incompatible demands for professionals, for example, because the time blocks might overlap and thereby disturb the emergence of connections.

Responding to Incompatible Demands: Work on It, Work Around It, Work Without It

As observations proceeded, I faced numerous situations in which the envisioned routine connections led to incompatible demands for participants. I further explored how professionals responded to these incompatible demands. From the data I derived three responses that routine participants developed to deal with these conflicting demands: *work on it*, *work around it* and *work without it*.

Work on It

The first response was labelled 'work on it'. This tag emphasizes that actors are 'busy doing things'. In the best way they can, they try to unite incompatible demands. The following vignette illustrates how one of the anaesthesiologists was confronted with conflicting demands. Because several delays occurred in the process, anaesthesia was demanded at two operating theatres at the same time.

We are halfway through the programme in the operating theatre[4] where four gynaecology operations are planned today. To resume the

---

[4]Field notes taken when shadowing a surgeon.

programme, the surgeon needs the anaesthesiologist for epidural anaesthesia and the time out. The assistant calls the anaesthesiologist to ask if he will come to the theatre for the time out. The anaesthesiologist answers that he is still very busy at the other theatre, where his task is complicated and will take a few more minutes. If they can wait a little longer, he will be there as soon as he can.

A few minutes pass by, in which the surgeon checks the clock several times. She sighs. 'Come on, hurry up! I have more to do today! And you know what, if the programme isn't finished in time, who has to inform the last patient that the surgery is postponed?! Me!' To the anaesthesia nurse: 'Can't you call one of the other anaesthesiologists? There might be someone wandering around, right?'

The anaesthesia nurse calls the staff room to see if someone is available. She hangs up the phone, and, satisfied, she says, 'There will be someone any minute!'

Again, a few minutes pass by. Then the second anaesthesiologist who was called enters the theatre and prepares for the epidural. Within seconds, the other anaesthesiologist enters the room. 'What are you doing here?' And then, annoyed, 'You should have called me if you didn't need me anymore. Now I have been working my ass off and rescheduled to be here, and for what? For nothing!'

The anaesthesiologist is not able to perform epidural anaesthesia in the two theatres at the same time. However, in the best way he can, he tries to manage these two processes anyway. This response involves informing the others to manage their expectations and prioritizing the different tasks. By giving priority to finishing the first task, the processes in the other operating theatre are put 'on hold'.

For the surgeon, this means that her series of routines gets disturbed. To keep the process going she tries to find a replacement for the anaesthesiologist, which again requires a lot of adjustment. In the end, the various professional routines seem to 'clash' rather than 'connect'. A conversation with the surgeon, later on, revealed some ideas about the ostensive dimension of the checklist routine. She argued that they were already used to performing safety checks before surgery, but with the formal checklist that requires all team members to be present, the process became more complicated and was often disturbed. In other words: 'It distracts me from what I'm doing'. So from a surgery perspective, the abstract idea of the checklist routine becomes a distraction, rather than a valuable tool. This ostensive idea did not come about in isolation, however; it was fuelled by the interrelation with other routines where a misfit occurred.

Because the different routines do not connect, the checklist not only seems to fall far short of expectations, but also seems to reinforce routines within the sub-disciplines—including senses of 'us' and 'them'—which makes the establishment of connections all the more difficult.

Work Around It

The second response reflects strategies used by professionals to get to the best result by adjustment; they work around (Morath and Turnbull 2005) the formal procedures. So rather than doing the best they can to make it work anyway, professionals fashion a solution to an unexpected problem or situation. This response has been identified in medical settings in earlier research (see e.g. Koppel et al. 2008).

Work arounds occurred in different ways. For example, they might involve completing and registering tasks at different moments than prescribed—surgeons who register the completion of the time out checklist before actually performing the checklist so they can move on more smoothly, or who perform the sign out checklist that entails recording post-operative agreements when these agreements are still to be made. Work arounds might also involve outsourcing operational tasks to someone else. The following vignette illustrates how an anaesthesiologists outsourced his tasks to a nurse anaesthesiologist who was lower in the hierarchy to deal with incompatible demands.

> The anaesthesiologist has been called because the patient is ready for the time out checklist. I follow the anaesthesiologist to the operating theatre,

but when we get there the surgeon is not present. The anaesthesiologist starts wandering around the surgery department to see if he can find the surgeon anywhere – without success. 'Okay then, I am going to do something else as well,' he says, apparently mostly to himself. To the operating assistant he says: 'Please call me when he returns.' We head back to the staff room. About ten minutes later the nurse assistant calls to inform us that the surgeon has returned and we can come for the time out.

At that time, however, we are already busy signing out in the other operating theatre. The anaesthesiologist asks the nurse anaesthetist to take over his tasks and says, 'You know the patient better than I do.'

In this situation, again an anaesthesiologist faced different care demands at the same time: a time out in one theatre and a sign out in the other. In order to not further delay the process, the anaesthesiologist decided to complete the task he was working on, and asked the nurse assistant in the other theatre to take over his tasks there.

During a coffee break later on, I asked the anaesthesiologist about this 'outsourcing'. He acknowledged that formally he was responsible and not allowed to delegate this work to someone lower in the hierarchy. However, trying to unite incompatible demands seemed unrealistic and thus unsafe, while this delegation seemed a reasonable option. The nurse anaesthesiologists are skilled, and they monitor the patient in the operating theatre the whole time, and therefore they do sometimes know the patient better than the anaesthesiologists. Moreover, they can always call for assistance. When I asked the anaesthesiologist if he felt uncomfortable with this situation he replied, 'That's why I made the call afterwards, just to be sure'.

This response comes out of the interrelation of routines in the first place, but is fuelled by the abstract idea of a routine that differs from the artefact. Although the artefact prescribes that anaesthesiologists have to fulfil these tasks themselves, they might feel that this is not necessary in order to deliver safe care. When routines are conflicting, they work around the formal procedure since they consider it safe.

Work Without It

The third response was labelled 'work without it'. With this response professionals did not strive to unite incompatible demands, but they explicitly made a choice. They prioritized one task over the other. This might mean working without the checklist, using it partly, or involving only a few team members. However, it might also mean working with the checklist and thereby casting aside another task, as the following vignette illustrates.

It is 7.50am on the day that I shadow one of the trauma surgeons. The day started at 7am with a round over the wards visiting the patients who are planned for surgery today or need extra care. We have to hurry to make it to the patient handover in the trauma surgery department where the status of the patients is discussed with all the trauma surgeons. The handover has already begun, and several clinicians are still walking in and out.

We have been at the handover for only five minutes when the trauma surgeon nods at me to leave. We have to go to the surgery department for the morning briefing. In the corridor I bump into the head of department; he argues that the idea of a briefing routine is highly valuable, but other routines have been overlooked. The morning handover has been a firmly established routine in the trauma surgery department, and the head of department underlines the value of discussing all the patients within the sub-discipline.

The introduction of the briefing, however, interfered with this routine since it requires surgeons to be at the operating theatre at 8am for the briefing. In order to manage this, they skip the handover. 'So they are going to a briefing to discuss the patients, but they haven't even properly discussed these patients within their own department,' he concludes.

The handover, a longstanding routine within the trauma surgery department, had been put into second place by the multidisciplinary briefing.

Professionals cannot fulfil these two tasks, and they prioritize the new routine. This made me wonder why they choose the new routine over the longstanding tradition.

Apparently, from a clinicians' perspective the ostensive dimension of the routine was that this briefing was 'important'. The briefing had been made into a formal routine and was reflected in several artefacts. In addition, surgeons argued that they were judged on their performance of the briefing—or rather, on the registration of the briefing. The patient handover in the trauma surgery department, although firmly institutionalized, was an informal routine. It was a longstanding tradition but was not backed by artefacts per se, and clinicians were not directly judged on it.

One routine; briefing, had been made more prominent, backgrounding the other, the handover. The new briefing routine thus partly replaced the existing handover routine. Performance measurement seems important for prioritizing routines.

Conclusion

This chapter has shown that a checklist in medical care does not stand on its own. Any routine is 'enmeshed in far-reaching, complex, tangled webs of interdependence' (Feldman and Pentland 2003, p. 104). I found the interdependence with conflicting routines to be an explanation for variability in routine performance. The routine connections as intended by the checklist are often not that straightforward and may even lead to incompatible demands for professionals. Rather than standardized responses, these incompatible demands require *responsiveness*. I derived three responses that professionals have developed to deal with incompatible demands: *work on it, work around it* and *work without it*. These responses often entail 'on the spot' decisions; there are no formal routines for prioritization.

The ethnographic data show how routine dynamics can be altered through the interaction of routines. For example, because of a conflict between existing routines and the checklist as an envisioned routine, ostensive aspects of the routine might change from a 'helpful tool' into 'a distraction' and thereby affect performance. How professionals value the checklist routine is thus not so much about the checklist itself, but about its (mis)fit with existing routines.

Furthermore, different groups (anaesthesiologists, surgeons) might have different understandings of a routine's ostensive aspect, e.g. what is important, what is the priority (cf. Zbaracki and Bergen 2010). When the checklist does generate a clash rather than a connection, this might also reinforce the strength of already existing routines within sub-disciplines and even result in conflict.

This analysis of the interrelation of routines highlights the importance of a different perspective on checklists in medical care. Thus far, checklists have predominantly been approached as instrumental coordination mechanisms, especially in implementation science. Routine theory underlines the importance of the interrelation with other routines, and provides a more contextual understanding. I conclude this chapter by claiming that checklists are actually 'hubs'. Checklists are about making connections between multiple professional routines. All these different routines, with their own structures, norms and values, have to connect in this hub. To get back to the Boeing that was considered 'too much airplane for one man to fly', I conclude this chapter by stating that in this case, it is not solely about too *many* processes for one checklist to capture, but about too many *different routines*. Because professional routines often fail to connect in a checklist, varying ostensive dimensions emerge— checklists therefore might lead to conflicts rather than connections. In order to make checklists routine practice in medical domains, attention should be paid to this interrelation with existing routines.

References

Abbott, A. (1988). *The system of profession: An essay on the division of expert labour.* Chicago: Chicago University Press.

Bosk, C. L., Dixon-Woods, M., Goeschel, C. A., & Pronovost, P. J. (2009). Reality check for checklists. *The Lancet, 374*(9688), 444–445.

Braham, D. L., Richardson, A. L., & Malik, I. S. (2014). Application of the WHO surgical safety checklist outside the operating theatre: Medicine can learn from surgery. *Clinical Medicine, 14*(5), 468–474.

Cabana, M. D., Rand, C. S., Powe, N. R., Wu, A. W., Wilson, M. H., Abboud, P. A. C., & Rubin, H. R. (1999). Why don't physicians follow clinical practice guidelines?: A framework for improvement. *JAMA*, *282*(15), 1458–1465.

Cruess, R. L., Cruess, S. R., Boudreau, J. D., Snell, L., & Steinert, Y. (2015). A schematic representation of the professional identity formation and socialization of medical students and residents: A guide for medical educators. *Academic Medicine*, *90*(6), 718–725.

Cyert, R. M., & March, J. G. (1963). *A behavioral theory of the firm*. Englewood Cliffs, NJ: Prentice-Hall.

Evetts, J. (2002). New directions in state and international professional occupations: Discretionary decision-making and acquired regulation. *Work, Employment & Society*, *16*(2), 341–353.

Feldman, M. S., & Pentland, B. T. (2003). Reconceptualizing organizational routines as a source of flexibility and change. *Administrative Science Quarterly*, *48*, 94–118.

Feldman, M. S., Pentland, B. T., D'Adderio, L., & Lazaric, N. (2016). Beyond routines as things: Introduction to the special issue on routine dynamics. *Organization Science*, *27*(3), 505–513.

Fourcade, A., Blache, J. L., Grenier, C., Bourgain, J. L., & Minvielle, E. (2011). Barriers to staff adoption of a surgical safety checklist. *BMJ Quality & Safety*, *21*(3), 191–197.

Freidson, E. (1994). *Professionalism reborn: Theory, prophecy, and policy*. Chicago: University of Chicago Press.

Gawande, A. (2010). *The checklist manifesto: How to get things right*. London: Picador.

Higginbottom, G., Pillay, J. J., & Boadu, N. Y. (2013). Guidance on performing focused ethnographies with an emphasis on healthcare research. *The Qualitative Report*, *18*(9), 1–6.

Hilligoss, B., & Cohen, M. D. (2011). Hospital handoffs as multifunctional situated routines: Implications for researchers and administrators. *Advanced Health Care Management*, *11*, 91–132.

Koppel, R., Wetterneck, T., Telles, J. L., & Karsh, B. T. (2008). Workarounds to barcode medication administration systems: Their occurrences, causes, and threats to patient safety. *Journal of the American Medical Informatics Association*, *15*(4), 408–423.

Levy, S. M., Senter, C. E., Hawkins, R. B., Zhao, J. Y., Doody, K., Kao, L. S., & Tsao, K. (2012). Implementing a surgical checklist: More than checking a box. *Surgery, 152*(3), 331–336.

McDonald, S. (2005). Studying actions in context: A qualitative shadowing method for organizational research. *Qualitative Research, 5*(4), 455–473.

Morath, J. M., & Turnbull, J. E. (2005). *To do no harm: Ensuring patient safety in health care organizations.* San Francisco: Jossey Bass.

Nelson, R. R., & Winter, S. G. (1982). *An evolutionary theory of economic change.* Cambridge, MA: Harvard University Press.

Pentland, B. T., & Feldman, M. S. (2008). Designing routines: On the folly of designing artifacts, while hoping for patterns of action. *Information and Organization, 18*(4), 235–250.

Pickering, S. P., Robertson, E. R., Griffin, D., Hadi, M., Morgan, L. J., Catchpole, K. C., & McCulloch, P. (2013). Compliance and use of the World Health Organization checklist in UK operating theatres. *British Journal of Surgery, 100*(12), 1664–1670.

Rydenfält, C., Johansson, G., Odenrick, P., Åkerman, K., & Larsson, P. A. (2013). Compliance with the WHO surgical safety checklist: Deviations and possible improvements. *International Journal for Quality in Health Care, 25*(2), 182–187.

Thomassen, Ø., Espeland, A., Søfteland, E., Lossius, H. M., Heltne, J. K., & Brattebø, G. (2011). Implementation of checklists in health care; learning from high-reliability organisations. *Scandinavian Journal of Trauma, Resuscitation and Emergency Medicine, 19*(1), 1.

Treadwell, J. R., Lucas, S., & Tsou, A. Y. (2013). Surgical checklists: A systematic review of impacts and implementation. *BMJ Quality & Safety, 23*, 299–318.

Tunis, S. R., Hayward, R. S., Wilson, M. C., Rubin, H. R., Bass, E. B., Johnston, M., & Steinberg, E. P. (1994). Internists' attitudes about clinical practice guidelines. *Annals of Internal Medicine, 120*(11), 956–963.

Urbach, D. R., Govindarajan, A., Saskin, R., Wilton, A. S., & Baxter, N. N. (2014). Introduction of surgical safety checklists in Ontario, Canada. *New England Journal of Medicine, 370*(11), 1029–1038.

Van Klei, W. A., Hoff, R. G., Van Aarnhem, E. E. H. L., Simmermacher, R. K. J., Regli, L. P. E., Kappen, T. H., & Peelen, L. M. (2012). Effects of the introduction of the WHO "Surgical Safety Checklist" on in-hospital mortality: A cohort study. *Annals of Surgery, 255*(1), 44–49.

Wallenburg, I., Hopmans, C. J., Buljac-Samardzic, M., den Hoed, P. T., & IJzermans, J. N. (2016). Repairing reforms and transforming professional practices: A mixed-methods analysis of surgical training reform. *Journal of Professions & Organization, 3*(1), 86–102.

Waring, J., & Bishop, S. (2013). McDonaldization or commercial re-stratification: Corporatization and the multimodal organisation of English doctors. *Social Science and Medicine, 82,* 147–155.

Zbaracki, M. J., & Bergen, M. (2010). When truces collapse: A longitudinal study of price-adjustment routines. *Organization Science, 21*(5), 955–972.

9

Sustaining Healthcare Service Improvements Without Collective Dialogue and Participation: A Route to Partial Failure?

Anne McBride and Miguel Martínez-Lucio

The state plays a pivotal role in reforming public sector workplaces in the UK (Bach and Kolins Givan 2012) but is only one of the social actors. Depending on the politics of the government in power, trade unions and professional associations also have some voice within these reforms, although the power to influence change is both invited and constrained. The most recent, and explicit, inclusion of trade unions in the public sector has been the large-scale pay modernisation in the NHS developed (only) in partnership by the 1997–2010 New Labour government (Bach and Kessler 2012). More likely is the crowding out of union involvement from negotiations over work allocation and work reorganisation (Carter et al. 2012; Clark 2014; Tailby et al. 2004). While this distancing of trade unions in the public sector workplace can be located within the tendency of national union leaders to focus on pay and job security, it

A. McBride (✉) · M. Martínez-Lucio
Alliance Manchester Business School, University of Manchester,
Manchester, UK
e-mail: a.mcbride@manchester.ac.uk

© The Author(s) 2018   **155**
A.M. McDermott et al. (eds.), *Managing Improvement in Healthcare*,
Organizational Behaviour in Health Care,
https://doi.org/10.1007/978-3-319-62235-4_9

also needs to be located within the motivations and processes of state-sponsored service and quality improvement initiatives. These initiatives have often focused on personalising services to the needs of 'customers' and establishing these needs as a counterbalance to the systems and content of collective dialogue (Kirkpatrick and Martínez Lucio 1995).

This twin-track approach to including *and* excluding public sector trade unions and professional associations (hereinafter also referred to as 'occupational collectivism') seems to mirror the fundamental contradictions referred to by Hyman (1987, pp. 30–35) in his classic intervention on management, whereby different functions of capital require both the coordination of complex operations and the disruption of labour power. Thus, we see managers engaging with occupational collectivism to enable the coordination of complex activities while at the same time disrupting the power of such collectives (by excluding them, for example) as a means of retaining control. For Hyman (1987, p. 30), there is no 'one best way' of harmonising or managing these contradictions, thus ensuring that any such attempts lead to 'partial failure'. While some academics (Townsend et al. 2014) apply Hyman to focus on the coordinating efforts of managers that lead to 'partial success', others (see Rubery et al. 2015) stress the 'partial failure' that lies in the more disruptive functions. This chapter does the latter and explores how managers may fail to develop effective service improvement teams because they are reluctant to create space for occupational collectivism. In the process of sending out mixed messages, managers risk losing the trust of workers and undermining the legitimacy of any changes.

This chapter explores these contradictions through the manner in which the state (primarily the New Labour governments but latterly the Conservative-led governments) has funded projects in NHS (England)[1] to improve service quality. Two aspects of state-sponsored service improvement interventions are examined in more detail. The first is the direct engagement of individual healthcare practitioners (rather than occupational collective groups) over issues of patient quality and safety. The second is the development of a cadre of workforce designers

---

[1]Bacon and Samuel (2017) provide evidence of different forms of union participation in NHS Scotland and Wales.

who become the *new* intermediaries between management and worker. Both interventions speak to the requirement of management to coordinate complex processes among healthcare practitioners but have been developed in the context of a policy discourse that is suspicious of healthcare practitioners, so that collective bodies have been effectively 'crowded out' from discussions of the healthcare labour process. While there is much discussion about healthcare professions resisting change (e.g. Currie et al. 2009) that might alarm managers, we argue that the absence of this collective voice has undermined the sustainability of any service improvement changes that require long-term workforce change. Such managerial actions ignore studies that indicate how new work practices can benefit from the legitimacy, knowledge and institutional support of trade unions (e.g. Ramirez et al. 2007; Bacon and Samuel 2017; Townsend et al. 2014). Indeed, these arguments have strong roots that are captured historically, globally, and by sector in a number of reviews (e.g. Wilkinson et al. 2010; Martínez Lucio 2013).

The presence of these 'co-ordinating' and 'disruptive' interventions is evident in the authors' analysis of three national government-funded initiatives that, in different ways, have sought to change work practices of healthcare practitioners as a means of service improvement. The first initiative is the Changing Workforce Programme (CWP) whereby the state sponsored role redesign across thirteen pilot sites in NHS (England) for subsequent adoption elsewhere (see Hyde et al. 2004 for further details). The second initiative is known as the 'Skills Escalator'. As detailed elsewhere, this was a concept to encourage organisational-wide workforce development around the needs of the patient (see McBride et al. 2006). It was also a core element of the Department of Health's *HR in the NHS Plan* (Department of Health 2002). The third initiative (Collaboration for Leadership in Applied Health Research and Care, CLAHRC) is a state-funded collaboration between clinicians and academics to encourage the greater take-up of evidence-based medicine and thereby close the gap between research and practice [see Harvey et al. (2011) and Hunt et al. (2016) for further details of the Greater Manchester (GM) CLAHRC featured in this chapter]. These initiatives (and underlying rationale) cover the period from 2002 to the present day.

At the heart of each initiative is a desire for service change and improvement, be it achieved through role redesign, workforce development or closer collaboration between researchers and practitioners. The rationale for these specific initiatives can be found in New Labour's ten-year programme of investment called *The NHS Plan* (Department of Health 2000) and its aspiration to provide a health service 'designed around the patient' (Department of Health 2000, p. 17). This desire, and accompanying discourse, continues with the 'new models of care' indicated in the *Five Year Forward View* (NHS England 2014) of the Conservative government of 2010. More recently, the House of Lords Select Committee on the *Long-term Sustainability of the NHS* identified service transformation as being 'at the heart of securing the long term future of the health and care systems' (Authority of the House of Lords 2017, p. 3).

This remainder of this chapter draws on the authors' studies of these three initiatives to illustrate the contradiction of directly engaging with individual healthcare practitioners and developing a new cadre of workforce designers while eschewing collective dialogue at the point of healthcare production. That the state has ongoing national-based relationships with trade unions and professional associations at the same time provides a further illustration of the contradictory nature of management (and of the dual role of the state as employer; Hyman 2008). We should not be surprised to find these managerial contradictions. What is new, however, is to examine them in the context of the perceived ongoing and potentially widening gaps in health, quality of care and funding (NHS England 2014). Viewing these contradictions in the context of limited service change leads to the chapter's conclusion that, without effective collective dialogue and participation, service improvements will be confined to small-scale changes enacted by individual healthcare professions with a particular interest in the change.

Direct Engagement with Individual Healthcare Practitioners

As indicated above, the desire to change the way the service was provided and which, and how, practitioners performed their roles was articulated in *The NHS Plan*. This document sent out a clear signal that it intended to improve productivity through the ending of 'old-fashioned demarcations between staff' and 'unnecessary boundaries... between staff' (Department of Health 2000, p. 27). This section indicates the manner in which the state directly engaged with individual healthcare practitioners to blur professional boundaries across these three initiatives.

The study of CWP indicates how healthcare practitioners were initially engaged in the process of redesigning roles by invitation to a CWP facilitated 'Role Redesign Workshop' at each pilot site. Instructions were for workshop participants to:

> ... generate ideas to improve the service through new ways of working... pick a small number (say four or five) of areas which are priorities in terms of high risk or greatest potential for benefit, and concentrate on these. (Hyde et al. 2004)

Role redesign appealed to participants and individuals were very positive about being involved in the CWP pilots (Hyde et al. 2004; McBride et al. 2005):

> I really did enjoy working with the therapists... There was not a day in that month when I... did not come away learning something. (Support Worker testing out new role)

> You start to become more determined yourself because you start realising that what you are actually arguing for is good and it would be a shame if it didn't happen. (Clinician testing out new role)

This desire of healthcare practitioners to tackle issues of concern to them can also be seen in the work of CLAHRC, which works directly with healthcare practitioners on service improvement ideas. For example, one stream of work of the GM CLAHRC has focused on improving the physical healthcare of patients with severe mental illness (SMI). In collaboration with local health service management, GM CLAHRC developed a programme of work that would enable the prevention, early diagnosis, treatment and management of physical health problems as part of the overall treatment and care of people with SMI (Hunt et al. 2016). This required direct engagement with a range of practitioners (e.g. General Practitioners (GPs), Practice Nurses, Care Coordinators, Support Workers, Assistant Practitioners, Nurse Trainers) regarding the development of a new role for a few individuals and a new set of responsibilities, or priorities, for most of the practitioners. Protected time was a key factor in the success of the new role (Hunt et al. 2016), but there is a danger that such protected time is viewed as evidence of the 'strong culturally conservative parts of our healthcare system, where the different professional tribes see particular ways of delivery services' (evidence from Director of Workforce at Department of Health to House of Lords' Select Committee, Authority of the House of Lords 2017). This chapter now moves to the new cadre of workers that has developed to direct and support service improvement work with clinicians.

Development of New Cadre of Workforce Designers/Service Improvement Specialists

CWP and GM CLAHRC provide examples of new roles explicitly created for service improvement. Starting with CWP, each pilot site was supported by two personnel. One was called a Workforce Designer, employed by the national CWP team to oversee the project and provide external links. The other was a Project Manager employed by the local organisation to work with staff on specific role redesigns. The following quotes convey the manner in which this new cadre of workers challenged current ways of working:

> People are very likely to say 'you can't do it' and it is just that it has always been a doctor who has done that…. things are done because they are always done that way and I challenge and say 'why' and they can't give you an answer, then they say 'that is just the way we do it'. (Hyde et al. 2004, p. 70)

> There were all sorts of challenges around the regulations on drugs. In fact when we tested it, and the project manager did a lot of work on this, there was no reason. There is no legislation that prevents that from happening… but it took us a long time to work through. (Hyde et al. 2004, p. 70)

CWP adopted a 'show us the evidence' approach to challenging current work practices, and by working directly with individual healthcare practitioners they stepped into quasi-managerial roles:

> CWP have been absolutely invaluable in getting past the 'little Hitlers', to coin a phrase. The big guns from the Department of Health come in and get it sorted. They say, 'unless it's against the law then you can't stop it'. (Person testing a new role) (Hyde et al. 2004, p. 36)

The cadre of personnel developed to work with practitioners within GM CLAHRC has a different approach, and this relates to their different starting point (Harvey et al. 2011). CWP could be viewed as a deliberate attempt to break down professional barriers and demarcations deemed to be detrimental to the patient and the delivery of efficient healthcare. In pursuit of healthcare modernisation, the CWP encouraged skill mix changes, the expansion or enrichment of jobs and the creation of wholly new jobs (McBride et al. 2005). The ultimate objective of GM CLAHRC is to increase the capacity of practitioners to put research evidence into practice, and as such it arguably falls within the remit of those agencies that Kirkpatrick et al. (2005, p. 92) identify as 'designed to put pressure on doctors to change their activities, in ways that managers themselves cannot undertake'.

The 'work' of GM CLAHRC has more immediate resonance with service improvement than the workforce modernisation agenda of CWP. The knowledge transfer associates (KTAs) appointed to the team (and employed by the NHS) work with University clinicians to understand the clinical evidence for particular practices (e.g. the importance of regular physical health checks of people for SMI) and work with NHS practitioners in the local context to facilitate the implementation of this evidence (Harvey et al. 2011).

Of interest to this chapter are those cases where the implementation of research evidence requires new ways of working or additional responsibilities. Using the aforementioned SMI project as an example, given the former infrequency of formal physical assessment of people with SMI within the primary, community and secondary care settings, the implementation of research (i.e. increasing the physical health assessment of this client group) will require practitioners to prioritise this among their list of tasks. In this case, the KTAs who have responsibility for facilitating an increase in physical health assessment become intermediaries between management and healthcare practitioners over issues of workload. The desire of the KTAs (and GM CLAHRC) to oversee a successful intervention inevitably has implications for workload discussions with management if workload is perceived as a barrier to embedding research into practice. In turn, this leads to KTAs being involved in quasi-negotiations with management as to how best to incorporate new tasks to ensure that this change is embedded in practice and sustainable. That this project becomes the channel for discussing workload issues illustrates the cumulative effect of a new cadre of workers directly engaging with healthcare practitioners such that collective dialogue is crowded out. In effect, a cadre of workforce designers and service improvement specialists have become the *new* intermediaries between employers and workers.

The Crowding Out of Collective Dialogue and Implications for Sustainability

While local CWP steering groups often included representation from professional bodies, such as the Royal College of Nursing, medical Royal Colleges, and/or trade union representatives, such collective groups were often discussed as being part of the 'problem' rather than having anything positive to contribute. For example, one role redesign was delayed because 'the staff group' ... 'wouldn't do it without remuneration' (Hyde et al. 2004, p. 706). The following quotation is from a CWP interviewee who noted that in one instance,

> ... the Royal College of Surgeons was incredibly unhelpful and started putting in 'you have to do this, that, and the other' which was actually regulating it to an extent that not even the doctors would be able to do it. (Hyde et al. 2004, p. 70)

However, such negative sentiments ignore the contribution that clinical groups *do* make to service improvement. Indeed, this tendency to blame occupational collectivism for resisting change continues to the present day and can be seen quite explicitly in the prominence given to the evidence of the 'conservative culture' in the Report of the House of Lords Select Committee (Authority of the House of Lords 2017, p. 38) and the absence from the report of evidence to the contrary (House of Lords Select Committee 2016).

The recognition of trade unions and professional groups was also mixed in the case studies undertaking Skills Escalator related activities. Generally, in line with national policy, the case study organisations engaged in joint union and management negotiations over a range of issues, but none appeared to have any influence over the manner in which the Skills Escalator concept was being applied (McBride and Mustchin 2007). Like their HR counterparts (McBride and Mustchin 2013), trade unions did not appear to have been deliberately excluded, but appeared to be crowded out by their prioritisation of other managerial issues and the existence and activities of other parties in this space. This has implications for the sustainability of service improvements.

Despite some successful developments that have led to considerable patient and practitioner outcomes, a number of service improvements have not spread beyond the individual practitioner, group or department that piloted the new way of working. For example, the national CWP team did not engage in discussions about pay, leaving this to the local organisation to settle. They used the phrase 'working differently, not working harder' to imply that increased pay would not necessarily follow the introduction of new ways of working, but it did mean that some staff groups disengaged from role redesign (Hyde et al. 2004). Likewise, some of the staff interviewed in the Skills Escalator project indicated that they did not feel they were being rewarded for the level of skill and experience they were bringing to their enhanced roles (Cox et al. 2008, p. 353). Research indicates that it was clinical managers who were most often involved in changing work practices, and that these changes were rarely broached through trade unions or professional associations (McBride and Mustchin 2007, 2013). With limited collective dialogue, and the crowding out of HR too (McBride and Mustchin 2013), the space in which to negotiate over the terms of such changes becomes non-existent. Change without apparent benefits can lead to discontent, as expressed in the words of one interviewee: 'if you're taking on extra skills … and you're not recognised for them too and not paid, it does sort of make you feel a bit demoralised' (McBride et al. 2006, p. 143). Indeed, the Skills Escalator interviews indicate the delicate balance to be achieved between developing staff, so that in the words of one interviewee '… you put more effort in your work and that you're qualified now, so you know, you don't say "oh, this will do" …' (McBride et al. 2006, p. 146), and demoralising staff because they believe they are doing the work of more qualified staff but feel this work is unrecognised because 'at the end of it all you're … still a health care [assistant]' (McBride et al 2006, p. 143).

Beyond healthcare, these quotations resonate with how the 1990s terminology of 'work smarter—not harder failed to translate into anything more than work intensification because of weaknesses in the regulation of employment' (Colling and Terry 2010, p. 19). By not including trade unions, there is limited discussion for determining how new tasks, or new roles, will be embedded as everyday components of a busy practitioner workload (McBride et al. 2005), which in turn undermines the benefits realisation of the ambitious pay modernisation developed in conjunction with trade unions (Buchan and Evans 2007).

Conclusions

This chapter has identified the manner in which the state has pursued its workforce modernisation agenda through a variety of service and quality improvement projects and initiatives. That this was a preferred choice is in keeping with the emphasis during New Labour (and subsequent governments) on changing employment relations in the absence of trade unions (Smith and Morton 2006) and through soft regulation (Stuart et al. 2011). The difference with this particular application in the NHS is that, rather than soft regulation being used to create new forms of employment relations in the vacuum of continued de-collectivisation, it appears to be encouraging de-collectivisation at the point of production in healthcare (where union density is still relatively high). We can see how despite the benefits to patient groups of particular quality improvements, trade unions and professional groups are becoming increasingly marginalised from collectively discussing, challenging and negotiating a number of changes to work practices which would benefit patients on a larger scale than those developed in pilot projects and individual studies.

Hyman (1987) argues that managerial contradictions can lead to 'partial failure', and we would argue that the absence of collective voice in service improvement is contributing to limited sustainability of what might actually be improved practices. If changes are not recognised and institutionalised through collective dialogue, we would argue that there is every opportunity for them to be edged out through the everyday use of worker discretion—not just through any hidden agenda by the professional, but because there has been no collective way of working out what is possible within already increasingly overcrowded workloads.

Acknowledgements This chapter is based on research gathered through three funded projects: an evaluation of the Changing Workforce Programme commissioned by the Department of Health, Research Policy Programme; a study of Skills Escalator activities commissioned by the Department of Health, Policy Research Programme; and the NIHR CLAHRC for Greater Manchester, funded by the NIHR and a number of primary care trusts in Greater Manchester. The views and opinions in the chapter do not necessarily reflect those of the NHS, the Department of Health or the NIHR.

References

Authority of the House of Lords. (2017). *The long-term sustainability of the NHS and adult social care*. HL Paper 151, select committee on the long-term sustainability of the NHS, Report of session 2016–2017.

Bach, S., & Kessler, I. (2012). *The modernisation of the public services and employee relations: Targeted change*. Basingstoke: Palgrave Macmillan.

Bach, S., & Kolins Givan, R. (2012). Varieties of new public management? The reform of public service employment relations in the UK and USA. *The International Journal of Human Resource Management, 22*(11), 2349–2366.

Bacon, N., & Samuel, P. (2017). Social partnership and political devolution in the National Health Service: Emergence, operation and outcomes. *Work, Employment and Society, 31*(1), 1–19.

Buchan, J., & Evans, D. (2007). *Realising the benefits? Assessing the implementation of agenda for change.* London: Kings Fund.

Carter, B., Danford, A., Howcroft, D., Richardson, H., Smith, A., & Taylor, P. (2012). 'Nothing gets done and no one knows why': PCS and workplace control of lean in HM revenue and customs. *Industrial Relations Journal, 43*(5), 416–432.

Clark, I. (2014). Health-care assistants, aspirations, frustration and job satisfaction in the workplace. *Industrial Relations Journal, 45*(4), 300–312.

Colling, T., & Terry, M. (2010). Work, the employment relationship and the field of industrial relations. In T. Colling & M. Terry (Eds.), *Industrial relations theory and practice* (3rd ed., pp. 3–25). Chichester: Wiley.

Cox, A., Grimshaw, D., Carroll, M., & McBride, A. (2008). Reshaping internal labour markets in the National Health Service: New prospects for pay and training for lower skilled service workers? *Human Resource Management Journal, 18*(4), 347–365.

Currie, G., Finn, R., & Martin, G. (2009). Professional competition and modernizing the clinical workforce in the NHS. *Work, Employment & Society, 23*(2), 267–284.

Department of Health. (2000). *The NHS plan.* London: Department of Health.

Department of Health. (2002). *HR in the NHS plan: More staff working differently.* London: Department of Health.

Harvey, G., Fitzgerald, L., Fielden, S., McBride, A., Waterman, H., Bamford, D., Kislov R., & Boaden, R. (2011). The NIHR collaboration for leadership in applied health research and care (CLAHRC) for greater Manchester: Combining empirical, theoretical and experiential evidence to design and evaluate a large-scale implementation strategy. *Implementation Science, 6,* 96.

House of Lords Select Committee. (2016). *Select committee on the long-term sustainability of the NHS, corrected oral evidence: The long-term sustainability of the NHS.* Session of Tuesday 1 November 2016, 10.10am. Retrieved from http://data.parliament.uk/writtenevidence/committeeevidence.svc/evidencedocument/nhs-sustainability-committee/longterm-sustainability-of-the-nhs/oral/43185.html. Accessed 11 Apr 2017.

Hunt, C., Spence, M., & McBride, A. (2016). The role of boundary spanners in delivering collaborative care: A mixed method evaluation. *BMC Family Practice, 17*(96), 1–10.

Hyde, P., McBride, A., Young, R., & Walshe, K. (2004). *A catalyst for change? The national evaluation of the changing workforce programme.* Manchester: University of Manchester.

Hyman, R. (1987). Strategy or structure? Capital, labour and control. *Work, Employment & Society, 1,* 25–55.

Hyman, R. (2008). The state in industrial relations. In P. Blyton, N. Bacon, J. Fiorito, & E. Heery (Eds.), *The Sage handbook of industrial relations* (pp. 258–283). London: Sage.

Kirkpatrick, I., & Martínez-Lucio, M. (1995). The politics of quality in the British public sector. In I. Kirkpatrick & M. Martínez-Lucio (Eds.), *The politics of quality and the management of change in the public sector.* Abingdon: Routledge.

Kirkpatrick, I., Ackroyd, S., & Walker, R. (2005). *The new managerialism and public service professions.* Basingstoke: Palgrave Macmillan.

Martínez-Lucio, M. (2013). Trade unions and organizational change in the public sector. In R. J. Burke, A. J. Noblet, & C. L. Cooper (Eds.), *Human resource management in the public sector* (pp. 236–251). Cheltenham: Edward Elgar.

McBride, A., Cox, A., Mustchin, S., Carroll, M., Hyde, P., Antonacopoulou, E., Walshe, K., & Woolnough, H. (2006). *Developing skills in the NHS.* Manchester: University of Manchester.

McBride, A., Hyde, P., Young, R., & Walshe, K. (2005). How the 'customer' influences the skills of the front-line worker. *Human Resource Management Journal, 15*(2), 35–49.

McBride, A., & Mustchin, S. (2007). Lifelong learning, partnership and modernization in the NHS. *The International Journal of Human Resource Management, 18*(9), 1608–1626.

McBride, A., & Mustchin, S. (2013). Crowded out? The capacity of HR to change healthcare work practices. *The International Journal of Human Resource Management, 24*(16), 3131–3145.

NHS England. (2014). *Five year forward view.* Retrieved from https://www.england.nhs.uk/wp-content/uploads/2014/10/5yfv-web.pdf. Accessed 11 Apr 2017.

Ramirez, M., Guy, F., & Beale, D. (2007). Contested resources: Unions, employers and the adoption of new work practices in the US and UK telecommunications. *British Journal of Industrial Relations, 45*(3), 495–517.

Rubery, J., Grimshaw, D., Hebson, G., & Ugarte, S. M. (2015). 'It's all about time': Time as contested terrain in the management and experience of domiciliary care work in England. *Human Resource Management, 54*(5), 753–772.

Smith, P., & Morton, G. (2006). Nine years of new labour: Neoliberalism and workers' rights. *British Journal of Industrial Relations, 44*(3), 401–420.

Stuart, M., Martínez-Lucio, M., & Robinson, A. (2011). 'Soft regulation' an the modernisation of employment relations under the British labour gov ernment (1997–2010): Partnership, workplace facilitation and trade unio change. *The International Journal of Human Resource Management, 22*(18 3794–3812.

Tailby, S., Richardson, M., Stewart, P., Danford, A., & Upchurch, N (2004). Partnership at work and worker participation: An NHS case stud *Industrial Relations Journal, 35*(5), 403–418.

Townsend, K., Wilkinson, A., & Burgess, J. (2014). Routes to partial suc cess: Collaborative employment relations and employee engagement. *Th International Journal of Human Resource Management, 25*(6), 915–930.

Wilkinson, A., Gollan, P. J., Marchington, M., & Lewin, D. (2010). *Th Oxford handbook of participation in organizations.* Oxford: Oxfor University Press.

10

Disseminating from the Centre to the Frontline: The Diffusion and Local Ownership of a National Health Policy Through the Use of Icons

David Greenfield, Margaret Banks, Anne Hogden
and Jeffrey Braithwaite

Introduction

Regulatory bodies, such as the Care Quality Commission in England, continually confront the issue of how to diffuse health policies effectively, particularly those directed at changing frontline clinical practice in support of quality and safety. For any policy, key questions

D. Greenfield (✉)
Australian Institute of Health Service Management, University of
Tasmania, NSW, Australia
e-mail: david.greenfield@utas.edu.au

M. Banks
Australian Commission on Safety and Quality in Health Care, Sydney,
Australia

A. Hogden · J. Braithwaite
Australian Institute of Health Innovation, Macquarie University, NSW,
Australia

© The Author(s) 2018
A.M. McDermott et al. (eds.), *Managing Improvement in Healthcare*,
Organizational Behaviour in Health Care,
https://doi.org/10.1007/978-3-319-62235-4_10

regulators struggle with are: how can they know the extent to which policy is disseminated to frontline clinicians; and how to embed and sustain improvements? Policy dissemination has proven effective in some circumstances. For example, the implementation of World Health Organisation surgical checklists has been both extensive, and shown to improve safety (Truran et al. 2011). However, many policy projects fail to achieve their goals (Kapsali 2011; Giguère et al. 2012) or their influence on clinical practice varies significantly (Giguère et al. 2012; Daniel et al. 2013; Prior et al. 2014). The challenge of policy dissemination is one facing all jurisdictions, in both developing (Khayatzadeh-Mahani et al. 2013; Kilewo 2015) and developed countries (Seddon et al. 2013; Lauvergeon et al. 2012). Furthermore, it encompasses every aspect of healthcare from organisational issues, including clinical governance (Khayatzadeh-Mahani et al. 2013) and health planning (Kilewo 2015), through to clinical service delivery, for example, maternity services (Rideout 2016), chronic disease management (Lauvergeon et al. 2012) and care coordination in aged care (Seddon et al. 2013).

Additionally, policy instruments may produce effects that are independent of their intended objectives (Clavier et al. 2012). In more extreme cases, policy implementation has contributed to incomplete process improvement, risks to patients and negative effects on staff (Weber et al. 2011). Being able to achieve ownership of policy at a local level, while ensuring a consistent national message, adds a further dimension to the issue. Knowing ahead of time which strategy is the most appropriate to facilitate dissemination and implementation is generally not readily available information to policymakers or senior executives sponsoring take up and spread (Brusamento et al. 2012; Grimshaw et al. 2004).

Healthcare is a complex adaptive system (Greenfield 2010). Healthcare is characterised by complicated, layered cultures and subcultures, nested behaviours, a multiplicity of provider roles, hierarchies and heterarchies, power and politics (Braithwaite et al. 2010). Simplistic linear models suggesting policy will be implemented unproblematically, will always meet these real world characteristics and fall short.

These challenges have raised awareness that the issues of dissemination and implementation need to be considered from the outset of policy formulation, and throughout its development (Kilewo 2015; NHMRC 2000), including identifying strategies, activities and practical tasks to promote uptake (Grimshaw et al. 2004; ANAO 2006; Lustria et al. 2013). However, policy diffusion is known to be a complex and uncertain undertaking. Networks facilitate the spread of information but context and interpretation play a significant role, with local environments shaping uptake and outcomes (Stone 2012). Policy uptake has been shown to be mediated by two-way communication between policymakers and those with responsibility for implementation. An issue largely unexamined is the impact which different forms of communication have on the movement and uptake of policies (Park et al. 2014). Policy outcomes, or changing behaviour at the frontline of healthcare, is one (or more) step further. When made available in the clinical environment, is the policy message sent, the message received? Central to this understanding is the idea of translation—that is, that policy has to be created anew when applied and appropriated into a specific context by an individual (Park et al. 2014). This requires learning, which viewed through Dunlop and Radaelli's (2013) typology of policy learning, involves deliberation or reflexive learning. Therein lies a key challenge: how to create space for frontline healthcare professionals for learning to take place?

Prompts and reminders are one component of an intervention strategy necessary for disseminating information and changing the behaviour of individuals and the settings in which they work (Giguère et al. 2012; Bywood et al. 2008). They are believed to be successful when embodying a clear and simple message, relevant to the task at hand and easy to use (Bywood et al. 2008). A communication strategy that targets and delivers the key message when needed, without the use of jargon, has been recommended (Shayo et al. 2014). Additionally, attention to and acceptance of policy is influenced by an organisational culture of reading policy and clinical material (Shayo et al. 2014).

The Australian Commission on Safety and Quality in Health Care (ACSQHC) has been responsible for the development and implementation of a national policy reform centred upon a new health service accreditation scheme and ten new associated standards. The ACSQHC describes the aim of the new National Safety and Quality Health Service (NSQHS) Standards 'to drive the implementation of safety and quality systems and improve the quality of health care in Australia. The 10 NSQHS Standards provide a nationally consistent statement about the level of care consumers can expect from health service organisations' (ACSQHC 2013). The NSQHS Standards were designed with coloured pictured icons to identify and brand each standard individually, and as a set (Fig. 10.1). Organisations are encouraged by the ACSQHC to use the icons as prompts and reminders in their education and preparation for accreditation activities and ongoing safety and quality initiatives. The ACSQHC has an established web-based approval process for

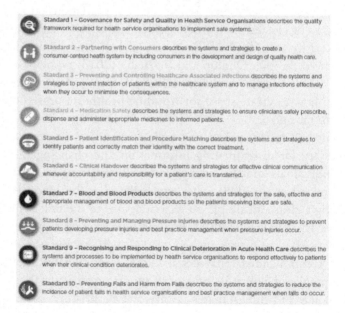

Fig. 10.1 The NSQHS Standards and icons (ACSQHC 2013). Reproduced in this book with permission from the Australian commission on safety and quality in health care

organisations to utilise the NSQHS Standards icons. To gain permission and access to the use of the icons, applicants are required to register their organisation and proposed use of the symbols.

In this study, we sought to answer the question: how can the ACSQHC know the extent to which the NSQHS Standards information is disseminated to frontline clinicians through the layers of a complex adaptive system, given the characteristics we describe? To do this we conducted a study examining the diffusion of the NSQHS Standards, via the use of the icons, across the Australian healthcare system.

Methods

The Accreditation Collaborative for the Conduct of Research, Evaluation and Designated Investigations through Teamwork (ACCREDIT) project has been investigating health service accreditation in Australia (Braithwaite et al. 2011). The ACCREDIT studies are informed by research conducted by study partners (Braithwaite et al. 2011; Greenfield and Braithwaite 2008; Greenfield et al. 2013; Greenfield et al. 2012a), including reviews of the healthcare accreditation literature (Greenfield and Braithwaite 2008; Greenfield et al. (2012b); Hinchcliff et al. 2012). Ethics approval for the study was given by the authors' university Human Research Ethics Committee, approval number: 10274.

Document analysis of two administrative databases collected and maintained by the ACSQHC was conducted. First, to identify the rate of diffusion of the NSQHS Standards via the use of icons across the health system, the administrative accreditation programme database was reviewed. This database records the organisations required to be accredited against the NSQHS Standards, and their progress in doing so. The review identified the total number of organisations and how many have been accredited against the new standards. Second, to consider the transmission and use of the NSQHS Standards icons within health services, the logo and icon database was audited. The database record was examined, line-by-line, to identify the characteristics of applicants and how and where they have used the icons. Analysis criteria were: locations and types of health organisations; organisational departments or settings; resources in which they are presented; and purpose(s) of use.

Findings

There are 1353 hospitals and day procedure services required to be accredited under the ACSQHC accreditation scheme against the NSQHS Standards; this includes all public and private hospitals in the country. The introduction of the scheme, transitioning from the previous programme, was planned to be completed over a 1-year period. The ACSQHC reports that all 1353 services will have been assessed against the NSQHS Standards by the end of 2015.

The diffusion of the NSQHS Standards and icons into health services commenced in 2012, to enable institutions to prepare for the new accreditation assessment requirements. By mid-2015, over 440 applications to use the icons had been received.

As Table 10.1 summarises, analysis of the records shows there is considerable dispersion and use of the icons. That is, they are used: in a variety of health organisations across all states and territories; in a range of settings within those organisations; and in numerous resources for multiple purposes. The icons are being used in all Australian states and territories (Table 10.2), by services across the health continuum, in both public and private sectors, and associated bodies, including health departments, accrediting agencies and professional associations (Table 10.3). Within institutions, the variety of departments or services using the icons ranged from policy and quality and safety units, training and promotional departments, to frontline clinical services and wards (Table 10.4). The icons were embedded into organisational documents, staff materials, educational and patient care resources and policy documents (Table 10.5). Icons served the dual purposes of promoting the NSQHS Standards and providing a visual reminder for staff of safety and quality responsibilities to patients. The vast majority of requests, or over 95%, were to use the complete set of ten icons. Submissions for the use of individual icons were the exception.

Table 10.1 Summary of the dispersion and use of the NSQHS Standards icons

Category	Details
Dispersion of icons	All Australian states and territories
	Organisations from the four most populous states accounted for 90% of the icon use
Organisations using the icons	State health departments
	Peak bodies and associations
	Public hospitals
	Private hospitals and day procedure centres
	Community health services
	Accreditation agencies
	Aged care services
	Publishing companies
Departments or services using the icons	Policy units
	Education and training departments
	Quality and safety units
	Clinical departments, services and wards
	Promotional departments
Resources where icons are embedded	Organisational documents—for example, strategic and operational plans, reports, toolkits, committee terms of reference and meeting minutes
	Staff materials—for example, posters, newsletters, intranet homepage, memos and email footers
	Education and training resources—for example, posters and presentations
	Patient care resources—for example, badges, t-shirts and magnets
	State and regional policy and procedure documents
Purpose of use	For branding or promotion of the NSQHS Standards
	As a visual reminder to staff of safety and quality responsibilities to patients

Table 10.2   State or Territory use of the NSQHS Standards icons

State/Territory	Count
Victoria	142
New South Wales	100
Queensland	68
Western Australia	66
South Australia	33
Tasmania	15
Australian Capital Territory	14
Northern Territory	7
Total	445

Table 10.3   Organisational use of the NSQHS Standards icons

Category	Count
Public hospital	201
Private hospital and day procedure centre	133
Community health service	79
State health department	38
Aged care service	32
Accreditation agency	7
Peak body and association	4
Publishing company	3
Total	497

Table 10.4   Departmental use of the NSQHS Standards icons

Category	Count
Quality and safety unit	159
Clinical department, service or ward	93
Education and training department	35
Policy unit	11
Promotional department	10
Total	308

Table 10.5   Resources location of the NSQHS Standards icons

Resource	Count
Organisational documents	292
Education and training resources	179
Staff materials	85
Patient care resources	32
Total	588

Discussion

The spread and adoption of the NSQHS Standards into health services has been able to be tracked through a novel strategy requiring registration for the use of icons. Hence, the ACSQHC has the capacity to identify the extent to which their policy is disseminated and made available to professional groups across the country. The figures reported represent the primary use of the icons. Applicants' secondary and further use of the icons is anecdotally reported but is not able to be measured; icon dissemination is wider and more diverse than the figures signify. Additionally, the process of registration could deter their uptake by some organisations or services within them.

The presentation of the NSQHS Standards in images, relevant to the clinical issue focused upon, has achieved several outcomes. First, as demonstrated in Fig. 10.1, the icons, individually and collectively, provide a simple visual representation and reminder of complex clinical issues, policy requirements and required staff behaviours. Second, they allow for the innovative distribution and promotion of NSQHS Standards in local contexts; this process encourages adoption and creativity in use, a requirement acknowledged as key to achieving uptake and sustaining improvement (Rubenstein et al. 2014). Third, uptake of the icons—across all Australian states and territories, types of health organisations, and used in a variety of settings and resources to achieve a range of purposes—demonstrates professionals' and organisations' recognition and acceptance of the safety and quality issues and NSQHS Standards. The icons have been and can be further used to promote the engagement of staff in quality and safety thinking and actions, and participation in the accreditation process. Organisations are known to use the icons in marketing and promotional activities to symbolically display their achievements and safety and quality credentials. The icons have been used to promote the uptake of the policy, addressing a key challenge associated with the dissemination of directed to frontline professionals (Grimshaw et al 2004; ANAO 2006; Lustria et al. 2013; Bywood et al. 2008).

This study has highlighted an effective policy communication strategy that can be used in any regulatory context. Icons and an associated database, which can be paper or electronically based as resource constraints allow, can be used to promote, track and assess the dissemination of policy and progress towards achieving its goals. This is of significant value as many projects lack such capability (Kapsali 2011; Prior et al. 2014; Shayo et al. 2014). The information allows policymakers or senior executives sponsoring the take-up and spread of policy to identify gaps in clinical areas or professional groups requiring additional targeted dissemination strategies (Brusamento et al. 2012; Lustria et al. 2013). Health workers can adapt the implementation of icons in line with their specific contexts, making their use locally appropriate while still maintaining a coherent national policy message. Communication in complex adaptive systems such as health care is challenging (Greenfield 2010; Braithwaite et al. 2010), but this study is evidence of the value of visual symbolic prompts as a component for disseminating information to frontline staff (Bywood et al. 2008). In settings with lower literacy levels the ability to make visible quality and safety requirements could be a significant factor in achieving positive patient outcomes (Shayo et al. 2014).

Part of the difficulty in seeking to answer the question of whether practice has changed in line with policy is that translation can be either coercive or voluntary (Park et al. 2014). At the organisational level there has been coercive translation of policy with the accreditation programme being mandatory, while at the frontline level of care the translation of policy into practice is voluntary. Individuals have to choose to take action, and identifying who has or has not is not always possible. Furthermore, evidence from research across the last two decades has confirmed that local context shapes the application and interpretation of policy (Stone 2012). Individuals within specific contexts are encouraged, mentored and required by systems, or not, to implement policy. While the argument has been made that the dissemination of the policy has been achieved, as evidenced by the database recording the uptake and adaption of the icons, the issue as to whether frontline practice has changed remains an open question. The visual representation of policy via icons is a simple strategy by which to bring, and reinforce, the quality and safety message into the frontline context. However, there are limitations with this information exchange. The pictorial communication is static, isolated and does not allow for exchange between the policymakers and frontline. This form of dissemination that effectively transfers a reminder into the local care context is also a barrier to increasing the chance of policy implementation success (Park et al. 2014). Does the message sent match the message received, or alternatively, did the icon create space for learning to take place? The database evidence shows that the icons are employed to promote awareness via reports, procedures and posters. However, frontline care contexts are not renowned for being circumstances in which staff have time and space to engage in reflexive learning with such items. On most occasions, the icon use will be self-directed, unstructured and judged as relevant, or not, by their own experience (Dunlop and Radaelli 2013). This means that even though it is possible to identify that the policy has diffused into frontline contexts, other assessments and interventions are necessary to determine if it is changing clinical practice.

Conclusion

There is evidence that the NSQHS Standards are increasingly embedded across and within the Australian healthcare system. The icons are being used as a visual stimulus signifying the quality and safety priorities for healthcare professionals; ripples of change continue to permeate through the health system via a tangible visual phenomenon. Icons are a strategy for promoting improvements in quality and safety that can be adopted and adapted to suit different regulatory settings.

The creative representation of policy in the form of icons has allowed for ease of distribution, uptake, presentation and recognition across a variety of organisational and clinical settings. The icons have proven to be an effective strategy for both the widespread diffusion and local ownership of a national health policy to those at the frontline of healthcare delivery.

References

ACSQHC. (2013). *Accreditation and the NSQHS Standards*. Retrieved from http://www.safetyandquality.gov.au/our-work/accreditation-and-the-nsqhs-standards/. Accessed 14 March 2016.

Australian National Audit Office. (2006). *Implementation of programme and policy initiatives: Making implementaiton matter*. Canberra: Commonwealth of Austalia.

Braithwaite, J., Greenfield, D., & Westbrook, M. T. (2010). Converging and diverging concepts in culture and climate research: Cultate or climure? In J. Braithwaite, P. Hyde, & C. Pope (Eds.), *Culture and climate in health care organisations* (pp. 7–18). Basingstoke: Palgrave-Macmillian.

Braithwaite, J., Westbrook, J., Johnston, B., Clark, S., Brandon, M., Banks, M., et al. (2011). Strengthening organizational performance through accreditation research: The ACCREDIT project. *BMC Research Notes, 4*, 390.

Brusamento, S., Legido-Quigley, H., Panteli, D., Turk, E., Knai, C., Saliba, V., et al. (2012). Assessing the effectiveness of strategies to implement clinical guidelines for the management of chronic diseases at primary care level in EU Member States: A systematic review. *Health Policy, 107*(2–3), 168–183.

Bywood, P., Lunnay, B., & Roche, A. (2008). *Effective dissemination: A systematic review of implementation strategies for the AOD field*. Adelaide: National Centre for Education and Training on Addiction.

Clavier, C., Gendron, S., Lamontagne, L., & Potvin, L. (2012). Understanding similarities in the local implementation of a healthy environment programme: Insights from policy studies. *Social Science and Medicine, 75*(1), 171–178.

Daniel, M., Kamani, T., El-Shunnar, S., Jaberoo, M.-C., Harrison, A., Yalamanchili, S., et al. (2013). National Institute for Clinical Excellence guidelines on the surgical management of otitis media with effusion: Are they being followed and have they changed practice? *International Journal of Pediatric Otorhinolaryngology, 77*(1), 54–58.

Dunlop, C. A., & Radaelli, C. M. (2013). Systematising policy learning: From monolith to dimensions. *Political Studies, 61*(3), 599–619.

Giguère, A., Légaré, F., Grimshaw, J., Turcotte, S., Fiander, M., Grudniewicz, A., et al. (2012). Printed educational materials: Effects on professional practice and healthcare outcomes. Cochrane Database of Systematic Reviews, 10. Retrieved from http://onlinelibrary.wiley.com/doi/10.1002/14651858.CD004398.pub3/abstract. Accessed 14 March 2016.

Greenfield, D. (2010). Accountability and transparency through the technologisation of practice. In J. Braithwaite, P. Hyd, & C. Pope (Eds.), *Culture and climate in health care organisations* (pp. 185–195). London: Palgrave Macmillan.

Greenfield, D., & Braithwaite, J. (2008). Health sector accreditation research: A systematic review. *International Journal for Quality in Health Care, 20*(3), 172–183.

Greenfield, D., Pawsey, M., & Braithwaite, J. (2012a). The role and impact of accreditation on the healthcare revolution (O papel e o impacto da acreditação na revolução da atenção à saúde). *Acreditação, 1*(2), 1–14.

Greenfield, D., Pawsey, M., Hinchcliff, R., Moldovan, M., & Braithwaite, J. (2012b). The standard of healthcare accreditation standards: A review of empirical research underpinning their development and impact. *BMC Health Service Research, 12*, 329.

Greenfield, D., Pawsey, M., & Braithwaite, J. (2013). Accreditation: A global regulatory mechanism to promote quality and safety. In W. Sollecito & J. Johnson (Eds.), *Continuous quality improvement in health care* (4th ed., pp. 513–531). New York: Jones and Barlett Learning.

Grimshaw, J., Thomas, R., MacLennan, G., Fraser, C., Ramsay, C., Vale, L., et al. (2004). Effectiveness and efficiency of guideline dissemination and implementation strategies. *Health Technology Assessment, 8*(6), 1–84.

Hinchcliff, R., Greenfield, D., Moldovan, M., Pawsey, M., Mumford, V., Westbrook, J., et al. (2012). Narrative synthesis of health service accreditation literature. *BMJ Quality and Safety, 21*, 979–991.

Kapsali, M. (2011). How to implement innovation policies through projects successfully. *Technovation, 31*(12), 615–626.

Khayatzadeh-Mahani, A., Nekoei-Moghadam, M., Esfandiari, A., Ramezani, F., & Parva, S. (2013). Clinical governance implementation: A developing country perspective. *Clinical Governance, 18*(3), 186–199.

Kilewo, E. G. (2015). Factors that hinder community participation in developing and implementing comprehensive council health plans in Manyoni District. *Tanzania. Global Health Action, 8*(1), 26461.

Lauvergeon, S., Burnand, B., & Peytremann-Bridevaux, I. (2012). Chronic disease management: A qualitative study investigating the barriers, facilitators and incentives perceived by Swiss healthcare stakeholders. *BMC Health Service Research, 12*(1), 1.

Lustria, M. L. A., Noar, S. M., Cortese, J., Van Stee, S. K., Glueckauf, R. L., & Lee, J. (2013). A meta-analysis of web-delivered tailored health behavior change interventions. *Journal of Health Communication, 18*(9), 1039–1069.

National Health and Medical Research Council. (2000). *How to put evidence into practice: Implementation and dissemination strategies.* Canberra: NHMRC.

Park, C., Wilding, M., & Chung, C. (2014). The importance of feedback: Policy transfer, translation and the role of communication. *Policy Studies, 35*(4), 397–412.

Prior, L., Wilson, J., Donnelly, M., Murphy, A. W., Smith, S. M., Byrne, M., et al. (2014). Translating policy into practice: A case study in the secondary prevention of coronary heart disease. *Health Expectations, 17*(2), 291–301.

Rideout, L. (2016). Nurses' perceptions of barriers and facilitators affecting the Shaken Baby Syndrome Education Initiative: An exploratory study of a Massachusetts public policy. *Journal of Trauma Nursing, 23*(3), 125–137.

Rubenstein, L., Khodyakov, D., Hempel, S., Danz, M., Salem-Schatz, S., Foy, R., et al. (2014). How can we recognize continuous quality improvement? *International Journal for Quality in Health Care, 26*(1), 6–15.

Seddon, D., Krayer, A., Robinson, C., Woods, B., & Tommis, Y. (2013). Care coordination: Translating policy into practice for older people. *Quality in Ageing and Older Adults, 14*(2), 81–92.

Shayo, E., Vaga, B., Moland, K. M., Kamuzora, P., & Blystad, A. (2014). Challenges of disseminating clinical practice guidelines in a weak health system: The case of HIV and infant feeding recommendations in Tanzania. *International Breastfeeding Journal, 9,* 188.

Stone, D. (2012). Transfer and translation of policy. *Policy Studies, 33*(6), 483–499.

Truran, P., Critchley, R. J., & Gilliam, A. (2011). Does using the WHO surgical checklist improve compliance to venous thromboembolism prophylaxis guidelines? *Surgery, 9*(6), 309–311.

Weber, E. J., Mason, S., Carter, A., & Hew, R. L. (2011). Emptying the corridors of shame: Organizational lessons from England's 4-hour emergency throughput target. *Annals of Emergency Medicine, 57*(2), 79–88.

11

Processes and Responsibilities for Knowledge Transfer and Mobilisation in Health Services Organisations in Wales

Emma Barnes, Alison Bullock and Wendy Warren

Introduction

Transferring and mobilising knowledge from research into healthcare delivery is an enduring international challenge (HM Treasury 2006; Mitton et al. 2007; Kitson et al. 2008). Research identifies better ways of providing healthcare or highlights mechanisms that no longer work, yet this knowledge often fails to influence the practices of those responsible for patient care. To inform decision-making in practice, research evidence needs to be 'available to those who may best use it, at the time it is needed ... in a format that facilitates its uptake', as well as 'comprehensible to potential users and ... relevant and usable in local contexts' (Sin 2008, p. 87). Finding ways to support access to knowledge that will help inform decisions is an important goal for health services research.

E. Barnes (✉) · A. Bullock · W. Warren
Cardiff Unit for Research and Evaluation in Medical and Dental Education (CUREMeDE), Cardiff University, Cardiff, UK
e-mail: barnesej@cardiff.ac.uk

© The Author(s) 2018
A.M. McDermott et al. (eds.), *Managing Improvement in Healthcare*,
Organizational Behaviour in Health Care,
https://doi.org/10.1007/978-3-319-62235-4_11

However, it cannot be assumed that presentation of the 'right research' will influence practitioners (Walshe and Davies 2013). Evidence use is a complex, social and dynamic process (Rushmer et al. 2015) involving 'the messy engagement of multiple players with diverse sources of knowledge' (Davies et al. 2008, p. 188). Davenport and Prusak (1998) explain how knowledge 'originates and is applied in the minds of knowers', and how in organisations 'it often becomes embedded not only in documents or repositories but also in organizational routines, processes, practices and norms' (p. 5). In an interactive model the linkage between researchers and research users is emphasised, and interpersonal exchange relationships are a means of bridging such knowledge gaps (Greenhalgh et al. 2004; Ward et al. 2009).

Collaborations have been established to link researchers, policymakers and service providers. In England, fifteen Academic Health Science Networks (AHSNs) were set up in 2014, with a focus on 'knowledge mobilization, rather than research production' (Walshe and Davies 2013). AHSNs bring most NHS organisations in England into collaboration with higher education institutions. Working alongside many AHSNs are Collaborations for Leadership in Applied Health Research and Care (CLAHRCs). Service-led and patient-focused, thirteen CLAHRCs aim to conduct high quality research, implement findings and increase NHS capacity. To facilitate knowledge mobilisation, many CLAHRCs have dedicated roles for translating and brokering knowledge.

The Scottish Executive and NHS Scotland has a team responsible for brokering activities including research mapping exercises, developing networks and communities of practice, and facilitating knowledge sharing events (Clark and Kelly 2005). They recommend using knowledge brokers as go-betweens, linking the policy, public sector, industry and academic communities (Scottish Government Knowledge Exchange Committee 2011).

In Wales, the Academic Health Science Collaboration (AHSC), formed in 2010, is a national programme with three regional entities in the South-West, South-East and North Wales. The AHSC identified knowledge transfer and mobilisation as a priority, and a national Task and Finish Group made recommendations on knowledge mobilisation policy (NISCHR AHSC 2014). The strategy of the South-East Wales

Academic Health Science Partnership (SEWAHSP) included a commitment to increase the speed and quality of 'translational' research and promote innovation in South-East Wales through strengthening collaborations between universities and NHS organisations.

The purpose of this study was to learn more about how knowledge was currently used to improve healthcare practice in Wales in order to better understand the difficulties and identify potential solutions.

Methods

The study employed qualitative interviews to explore opinions on the status of knowledge transfer and mobilisation (KT&M) within organisations, barriers and enablers and the potential of a knowledge broker role. The Research and Development (R&D) Directors in Health Boards across Wales with remit for KT&M (or their nominated representative) and Board Members of SEWAHSP (senior representatives from Health Boards, universities and other relevant organisations) were identified as key informants and invited to interview. We conducted 28 interviews, face-to-face at the participant's workplace or by telephone, utilising a semi-structured interview schedule which we sent ahead. Interviews typically lasted 30 to 45 minutes. All were audio-recorded, with permission. Audio recordings were transcribed and anonymised.

Research ethics approval was obtained from Cardiff University (REF/25.10.12). Research governance permission was acquired from participating Health Boards/Trusts.

We took a framework approach to the analysis (Ritchie and Spencer 1994). We developed a coding matrix of *a priori* themes based on Walker et al.'s (2007) four categories of factors that influence organisational change:

- *Context* factors in the external and internal environment
- *Content* the changes being transferred and implemented
- *Process* actions taken by the change agents
- *Individual dispositions* attitudes, behaviours, reactions to change

This model shares similarities with others (Kitson et al. 2008).

We also coded for understandings of knowledge mobilisation and whether KT&M processes were systematic. The coding process was iterative with identification of emergent subthemes. All coding decisions were discussed with the research team. The matrix allowed us to explore the analysis both across themes and across cases.

Results

Understandings of KT&M

Most interviewees thought KT&M was poorly defined. Interviewees expressed some confusion over the distinction between KT&M and other processes (such as audit, innovation, evidence-based practice, NICE guidelines or quality improvement).

> There's got to be a differentiation between R&D, KT, innovation – all these words are coming through at the moment, and they are confusing people. [Interview #20]

Some also suggested an understanding of knowledge which extended beyond formal sources; experiential forms of knowledge were valuable to decision-makers.

> There's a whole bunch of knowledge in an organisation that is not explicit ... that soft intelligence is very often not written down ... I would want to include that in part of the knowledge transfer process. [Interview #50]

Interviewees distinguished between the transfer of knowledge and its translation into practice, improved service delivery and patient outcomes.

> Basically we're talking about how does research really hit the ground to make a difference to people. [Interview #32]

The term 'knowledge transfer and mobilisation' was seen as useful for encapsulating both the transfer and implementation of knowledge.

Is KT&M Systematic?

Participants discussed the extent to which KT&M was embedded into practice within their organisation. While respondents indicated that it was an integral part of their personal professional practice, few saw it as an integral part of their organisation. KT&M activities tended to be ad hoc and individually driven, rather than embedded within organisations. Although some differences between professional groups, topic areas or improvement programmes were noted, the focus remained on individuals or teams:

> We still rely on individual teams to think about their own particular issues, their own particular services and where they might go to access evidence. [Interview #6]

Another interviewee explained how their organisation distributed newsletters and held dissemination meetings, but that these were ignored by most apart from those who were already research-focused ('the converted'). Information and knowledge sharing events for those in health service management roles were rarely mentioned.

However, practice differed by professional group, and national best practice guidelines and improvement programmes were said to have introduced a systematic process for some specialities:

> In terms of a specific technology in cancer, let's say a new drug, I think it's pretty well-developed. We all either have taken part in the clinical trials or we are contacted by the pharmaceutical company or NICE bring out a guidance – or it's in the press. [Interview #35]

However, the process was less straightforward for managers:

> We spend a lot of time talking about clinical evidence and research in relation to clinical care, but we don't spend so much time thinking about the evidence about the management of the service, the research into policy and practice that's around – how we deliver and manage and lead health and social care systems. [Interview #6]

1000 Lives Plus, a national NHS improvement strategy supported by Public Health Wales, was valued as a formalised technique for introducing service improvements.

> It actually introduced a structure by which evidence-based practice could be formally considered, discussions had about how we can change and implement it. [Interview #36]

Barriers to and Enablers of Knowledge Mobilisation

We asked participants what helped or hindered KT&M. In the analysis we coded these to the four factors in Walker et al.'s (2007) framework. These are summarised in Tables 11.1–11.4.

Context factors external to the organisation were thought to influence KT&M (Table 11.1). Positive government support for KT&M was said to be needed alongside policy linking social and health care, public health and universities. Some interviewees had observed a groundswell in KT&M policy in recent years. However, it was noted that a structured programme of support was also required to encourage and expect KT&M.

Interviewees argued against a one-size-fits-all approach, suggesting that approaches need to be adapted to local context. Within organisations, the culture and ethos, leadership and infrastructure (whether linkage was encouraged or whether silo-working dominated) were identified as influential factors. The pressure to deliver within a finite budget and extensive service demands could lead to a risk-averse culture. Lack of receptivity to new evidence, absence of an innovative culture and resistance to change were seen as barriers at all levels of the workforce. Participants highlighted the need for a supportive culture and a collegial approach within organisations. They remarked that culture change

Table 11.1 Context factors influencing KT&M

Barriers	Enablers	Illustrative extract
Competing priorities/agendas; meeting different demands on a finite budget	Targeted government policy to create a 'push' for change; policy based on meeting areas of patient need; research excellence framework giving attention to impact	Enabler: I think policy would be a good thing. Policy statement encouraging you, expecting it is an important thing to aid knowledge transfer. [#65]
Organisational culture which does not recognise the value of new evidence/change	Bottom-up changes in organisational culture to reframe professional role, valuing evidence and innovation; good leadership and management support at all levels—empowering staff and encouraging change	Enabler: Rather than top-down, if you encourage individuals to do it and to use their own skill and common sense to get information, I think that's a nicer way of doing it. [#17]
Unsupportive organisational infrastructure; no clear path for accessing/implementing evidence; reliance on personal interest or motivation	Clear signposting of opportunities/resources; support from an identified knowledge broker within the organisation	Barrier: I think a lot of the time it's a lack of receptivity, not a lack of enquiry or intelligence There's no system to it, and therefore people don't look for it. I think that's the challenge. [#35]
Lack of cross-professional working (professions, organisations, NHS and universities)	Multi-professional networks and face-to-face meetings; sharing knowledge and encouraging opportunities for innovation; engagement with organisations to make links (for example, SEWAHSP); communication	Barrier: Sometimes what you've got is professional tribalism … and that can be within professions and between professions. You've got hierarchies, it's a very difficult quagmire to find your way through. [#3]

comes from the ground up, and accordingly, staff members throughout the organisation need to be engaged in the process of change. However, frontline staff were considered to have limited opportunity for communicating successful changes to other departments.

Communication issues were discussed in terms of a lack of linkage between different sectors within and outside the organisation. Termed 'professional tribalism' by one interviewee, a lack of communication was noted within professions (staff hierarchy), between professions (nursing and medicine; clinicians and managers) and between organisations (primary and secondary care; NHS and universities). Creating networks and holding cross-disciplinary and multi-professional meetings was viewed as a way to help break down professional barriers, encourage communication between groups and facilitate organisations working as a whole.

The content or focus of the evidence was seen to impact on the mobilisation process (Table 11.2). Our participants wanted research to be relevant to population need, timely and motivating. Centring research on improving and addressing gaps in patient care was key. Alongside relevance for patients, having clear application to clinicians' practice was viewed as beneficial. 'Soft' intelligence and experiential knowledge were thought to be important in healthcare, yet they were not always considered legitimate by clinicians. One interviewee argued that the privileging of scientific knowledge in research excluded other types of knowledge and created distance between academic research and clinical practice.

The pressures of day-to-day work meant little time for reflection ('headroom') to consider the what, why and how of their current practice or to read new research (Table 11.3). Although important, interviewees suggested that KT&M was readily deprioritised when faced with day-to-day work pressures. Introducing supervision, coaching or feedback activities into routine practice was suggested as a way to tackle this, discussing the service and patient objectives and how they relate to their practice.

Participants commented that practitioners needed a coordinated approach since responding to different initiatives concurrently could be overwhelming. The need for collaborations and effective research/practice links was emphasised; stronger links between the NHS and universities were desired. While the importance of discussion was noted, it was

Table 11.2 Content factors influencing KT&M

Barriers	Enablers	Illustrative extract
Difficult to see relevance to practice in academic papers	Knowledge broker with good knowledge of target audiences to synthesise information and disseminate to appropriate professionals; involving NHS in research processes and researchers in dissemination; sharing examples of improvement arising from KT&M	Enabler: *It's about relevance. I think in terms of practitioners, staff nurses, ward sisters, community nurses, midwives on the ground, they've got to see that it is relevant for them and their practice and ultimately their client group and I think that bit is one of the challenges that people might find reading an academic paper.* [#32]
Valuing scientific research over organisational services research; 'soft' intelligence and experiential knowledge not valued as evidence	Recognising the importance of tacit knowledge/experience.	Enabler: *We need to reclaim some experiential knowledge.* [#65]

Table 11.3 Process factors influencing KT&M

Barriers	Enablers	Illustrative extract
Lack of time to reflect on practice/do KT&M activities	Embedding KT&M activities as part of every professional's role; protected time within workload	Barrier: *The pace of work is frenetic … very little thought about … what we actually do. I would like to see a more cerebral approach to health-care; where there's a bit more time to think.* [#1]
Overload of evidence; generalised dissemination of information; over-reliance on electronic dissemination (emails)	Dissemination of information that is timely, condensed, clinically rel-evant; central repository of relevant information	Barrier: *We are living in the middle of a knowledge explosion … The wrong thing to do is to be beating practi-tioners up because they haven't read enough papers, because they will never read enough papers.* [#3]
Overload of improvement initiatives	Focussed, targeted interventions/initia-tives aligned with local need; out-come measures in implementation programmes to provide guidance and reward achievement and belief in the process of change; manage-ment support	Enabler: *Through making a connec-tion with our universities and indus-try … to develop projects that could benefit patient care.* [#4]
Lack of communication; difficulty get-ting people together	Collaborations/partnerships and effec-tive research/practice links; greater cooperation between NHS and universities	Enabler: *Get people together and have discussion or somebody present about new research. It's seems to be a way that people pick up new ideas.* [#7]

Table 11.4 Individual factors influencing KT&M

Barriers	Enablers	Illustrative extract
'Inward-looking' staff members	The presence of 'can-doers'; outward-looking, motivated and open to change; leaders modelling good practice	Enabler: *The staff in the areas that are currently delivering … will sell it more with their nursing colleagues than me standing in front of them doing a bit of chalk and talk. So it's back to that ownership, and engagement and leadership. [#10]*
Lack of skills to appraise evidence	Embed skills in clinician education; knowledge brokers with research skills	Enabler: *It's important that we teach people the skills of appraising synthesised knowledge, and it's important that we commission synthesised knowledge. [#15]*

acknowledged that getting people together can be a challenge and communication via meetings sometimes resulted in superficial relationships. It was thought that more active and structured engagement was needed to develop deeper links.

It was acknowledged that an overwhelming amount of potentially relevant information is published and a targeted approach to accessing/disseminating is beneficial. They valued synthesised knowledge, with high-quality research filtered and summarised to capture the main relevance to managers/clinicians. Suggested enablers included making better use of librarians and R&D departments to access, assess and organise information that could be made more widely available or creating a central repository with summaries of evidence explaining how it relates to practice. Appropriate depth of information needed for different groups/problems was also discussed (sometimes providing just key messages, other times in-depth discussion).

Staff members' personal receptivity to KT&M was discussed (Table 11.4). Interviewees noted a lack of curiosity and motivation among some individuals to seek out new evidence. Conversely, the presence of 'can-doers' within the organisation, embracing change and championing KT&M, was seen as an enabler. These champions were believed to help challenge barriers, such as reluctance to change, by providing credibility, demonstrating investment and getting 'buy in'. The danger of relying too heavily on personality without a sustaining infrastructure was pointed out: the process needs to be embedded and stable enough to continue without their presence.

Who's Responsible for KT&M?

KT&M was seen as the professional responsibility of every practitioner, maintaining knowledge as a matter of patient safety.

> To me this is core stuff, it should be in all of their job descriptions. [Interview #5]

However, having nominated knowledge brokers within organisations was supported:

> I think you need to give somebody responsibility for the transfer of that knowledge, to ensure that when there is new evidence ... that it gets out to the right clinicians, and the right healthcare professionals, who can actually look to bring about the change and hopefully improve patient care. [Interview #20]

A knowledge broker's responsibilities were suggested as including collaborating with R&D and audit departments, building relationships with outside departments, identifying new research, disseminating it and observing outcomes. Such tasks were noted to already be part of the remit of HCRW registered research groups. Middle managers, directors, senior nurses or lead consultants were suggested for the role as such tasks were most closely aligned with their responsibilities. However, it was suggested that a potential risk of a nominated knowledge broker was that other professionals would pass all responsibility for KT&M to them. This highlighted the need for also embedding aspects of KT&M within all professional roles.

Discussion

The semi-structured nature of the interviews allowed participants to give their considered reflections. The use of existing frameworks (Walker et al. 2007) ensured a robust and consistent approach to data analysis. The findings verified the state of KT&M in Wales and the solutions needed to enhance progress as set out in the report of the KT Task and Finish Group (NISCHR AHSC 2014).

However, there were limitations. Although we accessed participants across Wales, we did not interview all R&D Directors or all SEWAHSP Board members. While we did not intend to formally assess knowledge mobilisation, a potential limitation is that the scope of the study did not allow us to verify participant's accounts of KT&M within their organisation.

Mindful of these limitations, a clear finding is that although there was interest in and appreciation of the value of knowledge mobilisation in Wales, processes were not systematic. Rather, they were reliant on individual interest and motivation. Compared to England, infrastructure targeting knowledge mobilisation is lacking, with no CLAHRC-style organisations in place. However, the HCRW AHSC identified KT as a priority and the national Task and Finish Group made recommendations on knowledge mobilisation policy (NISCHR AHSC 2014).

Barriers to knowledge mobilisation were like those noted in other research. Professionals' capacity to evaluate complex information was limited by time, means of accessing information and skills to distil implications for practice (Evans et al. 2013; Bullock et al. 2012; Baumbusch et al. 2008; Edwards et al. 2013; Golenko et al. 2012; Bullock et al. 2012). Relevance to practice influenced knowledge sharing activity, yet research may not address the current 'predominant concerns' in healthcare (Walshe and Davies 2013). Other studies have reported how managers source information, providing direct practical insight via informal interpersonal methods (Edwards et al. 2013; Dopson et al. 2013). Our participants discussed how disseminating evidence in timely, accessible formats and with clear relevance for practice would aid knowledge mobilisation. Their suggestions echoed others (Edwards et al. 2013) and included clear government policy linking knowledge mobilisation to R&D and quality improvement initiative which could help embed knowledge mobilisation in organisational practice.

In ever-changing systems, organisations need to be able to respond, learn and adapt (Schön 1973). Learning organisation theory explains the need to facilitate individuals' learning and link it to wider organisation achievement and practice change (Pedler et al. 1991; Senge 1990). Single-loop learning occurs where systems, values and goals remain unchallenged, whereas learning that explores systems and underlying assumptions is termed multi-looped learning. It is multi-loop learning and its outcomes that lead to organisational change (Argyris and Schön 1978). Systemic thinking within organisations allows individuals to see the long-term view of feedback (Senge 1990). Our findings show that a link from individuals to organisational change is missing, with learning remaining largely individually-motivated.

Making knowledge mobilisation work explicit and supported might consolidate KT&M as part of every professionals' role. Additionally, the knowledge mobilisation role of some team members could be optimised. Individuals skilled in appraising, synthesising and communicating knowledge to different target audiences could act as knowledge brokers. These brokers could aid networking, linking people with other relevant professionals and organisations—particularly those where there is little contact or trust (Ward et al. 2009; Bullock et al. 2016; Dobbins et al. 2009; Long et al. 2013; Williams 2002). Developing internal posts would foster the bottom-up change recommended by our participants. Knowledge mobilisation is embedded within complex organisational, policy and institutional contexts (Contandriopoulos et al. 2010), something which may challenge external boundary spanners (Evans and Scarborough 2014). Middle-managers in extended hybrid roles could bridge gaps between front-line employees and top-level management

(Birken et al. 2012; Burgess et al. 2015). However, their organisational ambidexterity may be impaired by professional demands and role conflict (Currie et al. 2015). This again underscores the need for clear organisational policy which values the broker role within a learning organisation. Care is needed so that these roles are seen as an adjunct rather than a replacement for personal knowledge mobilisation responsibility.

Conclusions

Whil we found awareness, interest and pockets of enlightened good practice in Wales, policy leadership is needed and a structured approach to ensuring that KT&M is an integral part of the day-to-day business of health organisations in clinical care. A systematic approach is needed to underscore the importance of KT&M and embed it in day-to-day activity.

Acknowledgements We gratefully acknowledge the support of Aneurin Bevan and Cwm Taf University Health Boards for funding this scoping review. We thank all the study participants who gave generously of their time. We are especially grateful to the input of our project reference group: Sue Bale, John Geen, Maureen Fallon, Rosamund Howell, Sue Figueirido and to Susan Denman for her critical feedback on earlier versions.

References

Argyris, C., & Schön, D. (1978). *Organizational learning: A theory of action perspective*. Reading, MA: Addison Wesley.

Baumbusch, J. L., Kirkham, S. R., Khan, K. B., et al. (2008). Pursuing common agendas: A collaborative model for knowledge translation between research and practice in clinical settings. *Research in Nursing & Health, 31*(2), 130–140.

Birken, S., Lee, S.-Y., & Weiner, B. (2012). Uncovering middle managers' role in healthcare innovation implementation. *Implementation Science, 7*(1), 28.

Bullock, A., Barnes, E., Morris, Z., et al. (2016). Getting the most out of knowledge and innovation transfer agents in health care: A qualitative study. *Health Service Delivery Research, 4*(33).

Bullock, A., Morris, Z., & Atwell, C. (2012). Collaboration between health service managers and researchers: Making a difference? *Journal of Health Service Research Policy, 17*(Suppl. 2), 2–10.

Burgess, N., Strauss, K., Currie, G., et al. (2015). Organizational ambidexterity and the hybrid middle manager: The case of patient safety in UK hospitals. *Human Resource Management, 54*(51), 87–109.

Clark, G., & Kelly, L. (2005). *New directions for knowledge transfer and knowledge brokerage in Scotland.* Edinburgh: Scottish Executive Social Research.

Contandriopoulos, D., Lemire, J. M., Denis, J. L., et al. (2010). Knowledge and exchange processes in organizations and policy arenas: A narrative systematic review of the literature. *Milbank Quarterly, 88*(4), 444–483.

Currie, G., Burgess, N., & Hayton, J. C. (2015). HR practices and knowledge brokering by hybrid middle managers in hospital settings: The influence of professional hierarchy. *Human Resource Management, 54*(5), 793–812.

Davenport, T., & Prusak, L. (1998). *Working knowledge: How organizations manage what they know.* Boston: Harvard Business School Press.

Davies, H., Nutley, S., & Walter, I. (2008). Why 'knowledge transfer' is misconceived for applied social research. *Journal of Health Service Research Policy, 13*(3), 188–190.

Dobbins, M., Robeson, P., Ciliska, D., et al. (2009). A description of a knowledge broker role implemented as part of a randomized controlled trial evaluating three knowledge translation strategies. *Implementation Science, 4*(1), 23.

Dopson, S., Bennett, C., Fitzgerald, L., et al. (2013). *Health care managers' access and use of management research* (Final report). Southampton: NIHR Service Delivery and Organisation Programme.

Edwards, C., Fox, R., Gillard, S., et al. (2013). *Explaining Health Manager' Information Seeking Behaviour and Use* (Final report). Southampton: NIHR Service Delivery and Organisation programme.

Evans, S., & Scarbrough, H. (2014, April). Supporting knowledge translation through collaborative translational research initiatives: 'Bridging' versus 'blurring' boundary-spanning approaches in the UK CLAHRC initiative. *Social Science & Medicine, 106,* 119–127.

Evans, B. A., Snooks, H., Howson, H., et al. (2013). How hard can it be to include research evidence and evaluation in local health policy implementation? Results from a mixed methods study. *Implementation Science, 8*(1), 17.

Golenko, X., Pager, S., & Holden, L. (2012). A thematic analysis of the role of the organisation in building allied health research capacity: A senior managers' perspective. *BMC Health Service Research, 12,* 276.

Greenhalgh, T., Robert, G., Macfarlane, F., et al. (2004). Diffusion of innovations in service organizations: Systematic review and recommendations. *Milbank Quarterly, 82*(4), 581–629.

HM Treasury. (2006). *A review of UK health research funding: Sir D Cooksey.* London: HM Treasury.

Kitson, A., Rycroft-Malone, J., Harvey, G., et al. (2008). Evaluating the successful implementation of evidence into practice using the PARiHS framework: Theoretical and practical challenges. *Implement Science, 3*(1), 1–12.

Long, J., Cunningham, F., & Braithwaite, J. (2013). Bridges, brokers and boundary spanners in collaborative networks: A systematic review. *BMC Health Service Research, 13*(1), 158.

Mitton, C., Adair, C., Mckenzie, E., et al. (2007). Knowledge transfer and exchange: Review and synthesis of the literature. *Milbank Quarterly, 85,* 729–768.

NISCHR AHSC. (2014). *Mobilising the use of research in practice for impacts on health and wealth—Recommendations of the AHSC Knowledge Transfer Task and Finish group to NISCHR, Welsh Government.* Cardiff: NISCHR AHSC.

Pedler, M., Burgoyne, J., & Boydell, T. (1991). *The learning company: A strategy for sustainable development.* London: McGraw-Hill.

Ritchie, J., & Spencer, L. (1994). Qualitative data analysis for applied policy research. In A. Bryman & R. Burgess (Eds.), *Analyzing qualitative research* (pp. 173–194). London: Routledge.

Rushmer, R. K., Cheetham, M., Cox, L., et al. (2015). Health Services and Delivery Research. In *Research utilisation and knowledge mobilisation in the commissioning and joint planning of public health interventions to reduce alcohol-related harms: A qualitative case design using a cocreation approach.* Southampton: NIHR. Journals Library.

Schön, D. A. (1973). *Beyond the stable state. Public and private learning in a changing society.* Harmondsworth: Penguin.

Scottish Government Knowledge Exchange Committee. Main Research Providers (MRPS). (2011). *Scottish Government Research Programme 2011–2016. Knowledge Transfer/Exchange (KTE) Strategy.* Edinburgh.

Senge, P. M. (1990). *The fifth discipline: The art and practice of the learning organization*. London: Random House.

Sin, C. (2008). The role of intermediaries in getting evidence into policy and practice: Some useful lessons from examining consultancy-client relationships. *Evidence & Policy, 4*(1), 85–103.

Walker, H. J., Armenakis, A. A., & Bernerth, J. B. (2007). Factors influencing organizational change efforts: An integrative investigation of change content, context, process and individual differences. *Journal of Organisational Change Management, 20*(6), 761–773.

Walshe, K., & Davies, H. T. (2013). Health research, development and innovation in England from 1988 to 2013: From research production to knowledge mobilization. *Journal of Health Service Research Policy, 18*(Suppl. 3), 1–12.

Ward, V., House, A., & Hamer, S. (2009). Knowledge brokering: The missing link in the evidence to action chain? *Evidence & Policy, 5*(3), 267–279.

Williams, P. (2002). The competent boundary spanner. *Public Administration, 80*(1), 103–124.

12

Accelerating Research Translation in Healthcare: The Australian Approach

Helen Dickinson and Jean Ledger

Introduction

Health systems are currently faced with a series of challenges, including the need to contain costs while maintaining quality, producing more seamless, co-produced services with users and making use of rapidly advancing technologies and innovations. The combination of these factors are playing out in different ways across national settings, but one challenge faced by all health systems is how to mobilize research and best practice knowledge in order to drive effective and efficient health services. During the 1990s we saw the rise of Evidence-Based Medicine (EBM) as an international paradigm, followed in the 2000s

H. Dickinson (✉)
University of New South Wales, Canberra, Australia
e-mail: h.dickinson@adfa.edu.au

J. Ledger
University College London, London, UK

© The Author(s) 2018
A.M. McDermott et al. (eds.), *Managing Improvement in Healthcare,*
Organizational Behaviour in Health Care,
https://doi.org/10.1007/978-3-319-62235-4_12

by academic and policy interest in research translation and impact, now core themes across many different policy areas and strongly influencing the healthcare sector. Despite keen interest in the EBM movement, however, there are still significant challenges in developing organizational arrangements that enable translational activity in practice.

Within the Australian context, a government study found that while the country compares well with the USA, Canada and a number of Western European countries in terms of the numbers of cited research papers, it places ninth for quality of scientific institutions, 72nd for innovation efficiency (innovation output relative to input) and less than one in two Australian businesses report innovative activity of any kind (Commonwealth of Australia 2015a). It concluded that 'new knowledge in itself is not enough to catalyse broad-based change across an economy' (ibid., p. iv), suggesting that research production alone is insufficient for driving system level change and knowledge mobilization.

A problem for funding agencies and policymakers tackling this issue has been the lack of evidence on the impact and effectiveness of particular mechanisms and strategies that support the active uptake of research into practice, especially at the national level (Tetroe et al. 2008, p. 150). Contributions from the field of biomedicine are helpful in this respect, having extensively mapped out the various different bottlenecks in the journey from scientific research and evidence to practice (Meslin et al. 2013). Khoury et al. (2007) argue there are distinct gaps along a research translation continuum which involves a spectrum of activities from basic scientific research to the development of evidence-based guidelines, the implementation of guidelines into healthcare practice and evaluation of outcomes. Each translation phase requires investment to effectively mobilize scientific knowledge for population benefit (Khoury et al. 2010): basic scientific discovery (T1) and clinical trials to produce evidence-based guidelines (T2); the development of evidence-based interventions appropriate for use in daily practice and implemented for patient benefit (T3); and evaluations of the impact of interventions with respect to population health outcomes (T4). To take one example with potential to impact on public health, the authors observe that in the final phase of a research translation journey into prostate-specific antigens, 'T4 research would involve looking at the health impact of prostate screening in unselected popu-lations (or real world practice)' (ibid., p. 519). This confirms that the journey from research and evidence to practice is not straightforward and requires resources and capabilities to overcome 'translational blocks' where knowledge and communications frequently become delayed (Thornicroft et al. 2011, p. 2016). Consequently, the research transla-tion journey has a complex link to research infrastructure that goes well beyond the impact of different forms of evidence-based guidelines on healthcare professionals' decision-making processes. Knowledge mobili-zation for population health benefit requires active interventions at dif-ferent institutional, organizational and occupational levels to support the movement of new knowledge into everyday practice.

In considering how to drive translational research in healthcare, the three groups that typically receive most attention are universities, healthcare organizations and clinical practitioners. However, indus-try remains a significant actor for supporting national innovation sys-tems and translating academic research into commercial products, so concepts such as the 'Triple Helix'—which stresses the value of rela-tional linkages across institutions and sectors for economic impact and growth—bring attention to partnerships between universities, govern-ment and industry in research translation processes (Etzkowitz and Leydesdorff 2000). An important feature of the Triple Helix concept is that it views institutional relations as evolving rather than fixed; 'knowl-edge flows' between separate academic, industrial and governmental spheres are not considered linear—as in from origin to end applica-tion—but as relational and grounded in historical patterns of interac-tion that can be reconstructed (Etzkowitz and Leydesdorff 2000).

At the organizational level, research on networks and collaborations encourage thinking about how knowledge exchange and innovation can be supported through alliances made up of diverse organizational members, including from the private and public sectors (Pittaway et al. 2004). Knowledge exchanged through network structures may bring about enhanced performance and economic advantages that cut across traditional sector divides; however, the characteristics of networks vary hugely. In the UK, research on mandated health networks suggests that network configurations can be highly adaptive and display different features over time (such as brokerage) that support the embedding of knowledge within practice (D'Andreta and Scarbrough 2016).

Against this theoretical research background, a range of different knowledge mobilization approaches have emerged that encourage the uptake of research evidence in practice: the use of academic practitioners to bridge health and research boundaries; tools for research translation (e.g. the Canadian Foundation for Health Improvement's Research Self-Assessment Tool (SAT)); and, in the UK, different organizational and structural arrangements such as Collaborations for Leadership in Applied Health Research and Care (CLAHRCs), Academic Health Science Centres (AHSCs) and Researcher-in-Residence models (Rowley et al. 2012; French et al. 2014). Australia, too, has a new and increasing focus at the federal level on the impact and commercialisation of research which looks set to give greater legitimacy to new type of organizational forms (National Health and Medical Research Council 2015). In this chapter, we map out the different arrangements and consider their key features and how they are intended to operate in practice. We focus on three particular approaches—Centres of Research Excellence, Advanced Health Research Translation Centres and Clinical Networks—exploring their purposes and potential to improve knowledge mobilization and research translation. We conclude by noting that the evidence base for these organizational arrangements is still developing, and that an overall emphasis on governance and structural arrangements may overlook significant processes of culture change and collaboration occurring locally.

The Australian Healthcare System

The Australian healthcare system has been described as 'one of the most fragmented health systems in the world' (Brooks 2011) with responsibilities split between different levels of government (local, state/territory, federal/commonwealth), as well as non-government sectors. Australia has universal healthcare through the Medicare scheme, but many people access private healthcare services either directly or via insurance schemes. By world standards, Australia has a 'good health system for reasonable per capita health expenditure' (McKeon et al. 2013, p. 9).

Yet, arguably, more can be done to ensure that the best research evidence is mobilized to help the system respond effectively to the challenges we will see in the coming years. In particular, the fragmentation and complexity of the system poses challenges for patients with complex and chronic diseases, and should the system remain the same, it is likely it will become more inefficient in dealing with rising rates of complex and chronic disease (Commonwealth of Australia 2015b).

The National Health and Medical Research Council (NHMRC) is an expert body that promotes the development and maintenance of public and individual health standards. The NHMRC is an independent statutory agency within the portfolio of the Australian Government Minister for Health and Ageing. The NHMRC has five priority actions to 'build a healthy Australia' (as set out in the NHMRC Strategic Plan 2013–2015). For the purposes of this paper, two of note are 'to accelerate research translation' and 'build Australia's future capability for research and translation'. As Australia's peak funding body for medical research, the NHMRC draws upon the resources of all components of the health system, including governments, medical practitioners, nurses and allied health professionals, researchers, teaching and research institutions, public and private programme managers, service administrators, community health organizations, social health researchers and consumers. Although universities sit outside the formal arrangements of the Australian health system, they provide invaluable support and training for healthcare professionals and researchers. Universities also provide collaborations to move research into practice, although this is left to the local level with different approaches in place around the country. The acceleration of translation of research in the Australian health system is a Priority Action Area of the NHMRC's Strategic Plan 2013–2015, with the aim of 'promot[ing] and accelerat[ing] the translation of evidence from research into improved care for patients, and thereby support[ing] a self-improving health system', as well as the 'translation of research into improved health policy and practice, and commercialisation' (National Health and Medical Research Council 2010). The NHMRC discussion paper envisaged 'universities, medical research institutes and hospitals working together to support research and research translation'. This was supported by calls to 'get competitive' internationally from the Deans of Australia's Group of Eight faculties of medicine (Fisk et al. 2011).

Having set out an overview of the challenges of the Australian health system and the role of the NHMRC, the following section moves on to consider the different institutional mechanisms that have been used to encourage and accelerate research translation.

Centres of Research Excellence

Centres of Research Excellence are relatively traditional research translation mechanisms aiming to provide support for university researchers to pursue collaborative research and develop capacity in research. These centres do not have a specific mandate around knowledge mobilization or research translation, although an increasing number are developing programmes that do this, possibly influenced by pressures for universities to demonstrate their impact beyond the academic sphere (Penfield et al. 2013). For example, an objective for these centres is to 'have an impact on the wider community through interaction with higher education institutes, governments, industry and the private and non-profit sector' (Australian Research Council 2015). As such, universities increasingly have a major role to play in developing solutions to healthcare and wider policy problems related to human populations, social inequalities and economic growth—extending their reach beyond traditional teaching and research functions. A particular concern in the Australian context is the diagnosis of 'a poor management culture of innovation and collaboration, and shortages in a range of skills... Australia is primarily a nation of adopters and modifiers operating behind the innovation frontier (and) ... should place more emphasis on improving levels of industry-research collaboration ... as first steps towards becoming a global leader in innovation. Collaboration between research and industry is one of the lowest in the OECD' (Office of the Chief Economist 2014). In this context, the mechanism of creating prestigious centres of research excellence with clear collaborative missions is seen as one way to improve these efforts.

Advanced Health Research and Translation Centres

Internationally, Academic Health Science Centres (AHSCs) are partnerships designed to accelerate research translation by integrating biomedical research, professional education and clinical care (Fischer et al. 2013). As major knowledge transfer endeavours, these can be viewed as 'academic' or 'clinical' 'enterprise organizations' (Ovseiko et al. 2014). The internal social and organizational processes of knowledge mobilization within AHSCs are not well understood, however (French et al. 2014). They differ significantly in internal clinical-managerial arrangements (Ovseiko et al. 2014), which may influence the types of knowledge mobilization strategies and institutional incentives employed. French et al. (2014, p. 389) caution that a particular organizational model 'does not necessarily determine whether or not an AHSC is successful in achieving its tripartite mission' of research, education and healthcare delivery.

The NHMRC's initiative to establish four Advanced Health Research and Translation Centres in Australia, based on the AHSC model, is intended to boost Australia's ability to compete internationally as leaders in healthcare research and education, and its translation into patient care. Although the relatively newly commissioned AHRTCs are in their infancy, some observers comment that determining appropriate models of governance is a major concern (Brooks 2011, 2009), echoing other international jurisdictions (Ovseiko et al. 2014; Davies et al. 2010). Within the Australian context, a potential conflict may occur when there are a number of institutes and organizations that function with a considerable amount of autonomy, and what each organization will have to 'cede' in order for the larger AHRTC to function (Brooks 2011). In order for these to succeed, governance change will be required with both 'top-down' and 'bottom-up' approaches to unite the many different entities in the AHRTCs towards 'a single mission' (Fisk et al. 2011). With these different models of governance also come diverse and different measures of success, and within the healthcare context not all players have community health outcomes as key drivers for success (especially with the increasing commercialization of the Australian healthcare context). This is especially noteworthy when one considers that the private sector is responsible for more than half of all healthcare delivery in Australia (Jennings and Walsh 2013).

Fisk et al. (2011) highlight key barriers to implementing AHSC or AHRTCs in Australia, including: turf wars between universities and hospitals over their diverse missions, priorities, operational frameworks and employment conditions, with process and contracts frustrating attempts to bridge the gap; the fact that three ASHC pillars are overseen by three separate federal government departments (and ministers); and the fact that additional players in the research sector—the independent medical research institutes—although affiliated with universities and tertiary hospitals, have at times eschewed translational links with clinical medicine in favour of basic science. Despite these concerns, there has been little research done to highlight the potential barriers that may prevent successful development of AHRTCs, or what mechanisms might facilitate their success.

Clinical Networks

Like Centres of Research Excellence and AHRTCs, clinical networks are also a mechanism that will be familiar to those in other jurisdictions, featuring in many different health systems around the world (Perri et al. 2006). The need for clinical networks emerged from a context in which the delivery of healthcare was becoming more specialist and the need for collaboration and cooperation between professionals ever more important when dealing with complex and 'wicked' problems (Thomas 2003; Ferlie et al. 2013). Alongside the challenges posed by the boundaries of traditional professions we have seen rapidly developing technologies, new knowledge and rising consumer expectations. Managed clinical networks have been viewed as one way of overcoming issues of fragmentation in this context, thereby improving patient care and also the efficiency and effectiveness of services (CanNet National Support and Evaluation Service 2008).

Clinical networks operate as a 'virtual team', where a group of healthcare professionals from different backgrounds work to deal with a particular problem. Depending on the problem being addressed this might include primary, secondary or tertiary care across different geographical areas. They can cover a particular disease specialty, function or location (Addicott et al. 2007). Australia has clinical networks in place across most states and territories in relation to a variety of different areas. Their precise makeup varies according to the different jurisdictions and local needs, but overall they are fairly well embedded within the Australian health system. Typically, one of the main aims of these mechanisms is to encourage the use of evidence and translation of knowledge into practice. One of their aims is to 'encourage best practice and improve access to teaching and research' (Government of South Australia 2007, p. 5). Usually, each has a chair (a professional who leads the network), some administrative and project support, a steering committee and may involve consumers. The evidence base for these mechanisms is overall mixed, although it suggests generally that when well managed according to a key set of priorities, clinical networks can be effective in driving the improvement of clinical care (Perri et al. 2006).

Discussion

Compared to other developed countries, Australia's development of formalized institutions and organizational structures for research translation has been more recent. This would seem to be somewhat unexpected given the complexity of the Australian healthcare system. As these examples of research translation organizations illustrate, a significant amount of future research investment seems likely to be focused in developing translation approaches. To date there is little evidence about how these mechanisms operate in an Australian context, although lessons are available from experiences in other jurisdictions. We now turn to this evidence and what it might tell us about how these mechanisms could operate in an Australian context. Broadly, what this evidence reveals is that a focus on formalized structures alone will not drive more effective research translation in practice. These are a necessary but insufficient contribution in mobilizing research translation and evidence use. Attention also needs to be paid to the relationships between key actors involved in the research translation continuum.

The more traditional approaches of Centres of Research Excellence, while important in generating new research, have less focus on translating evidence into practice. This is not specifically their role, although some are developing this translation aspect more organically through their particular local interests and in response to the shifting policy context. Clinical networks are a more recently established mechanism and a growing evidence base has emerged in recent years as to their impact (D'Andreta and Scarbrough 2016). A number of positive effects have been associated with these mechanisms including the interchange of evidence and ideas and the better use of resources (CanNet National Support and Evaluation Service 2008). There is an important caveat in the degree to which we see these impacts emerging from managed network arrangements. As Addicott et al. (2007) demonstrate in their study of managed clinical networks in the UK, if the broader context is not conducive to the operation of the network then it will not be successful. If they are simply seen as a structural panacea to implement nationally, it is unlikely they will succeed; networks take time to become established and need professionals to drive their internal processes (Ferlie et al. 2011).

Recent studies of AHSCs emphasize structural issues of organizational form, regulation and accountability (Ovseiko et al. 2014), while French et al. (2014) point to the importance of organizational and performance arrangements, including competing institutional pressures and interactions. Indeed, Fischer et al. (2013) study of the early organizational development of a large AHSC found that strong engagement and commitment by senior professionals (rather than formal organizational leaders or policymakers) played a key role in producing major institutional change, mobilizing emotion, values and collective affect. While the evidence base for these organizational arrangements is still developing, we note an overall emphasis on governance and infrastructure, rather than much focus on how these arrangements might develop locally significant capacity for research collaboration and translation. Without a more finely grained focus on how they serve models of research translation, it is difficult to derive lessons for developing such processes more systematically. What is clear from the evidence base is that we lack, overall, a good picture of the operation of these mechanisms. As such there is a lack of clarity about how these different mechanisms will operate in practice and how the entire knowledge mobilization journey will operate.

Across many different schools of thought, there is agreement that research evidence is insufficiently used in the practice of everyday healthcare, but there is rather less agreement over the processes that help in translating research and getting evidence into everyday use. Smith (2013) charts the emergence of ideas relevant to research translation and the variety of ideas that underpin the relationship between evidence and policy. She argues it is important we consider ideas of how research or evidence is translated into another sector, because this has a bearing on the types of mechanisms that are put in place to support these processes. What she is suggesting here is that we need to unpack assumptions about how evidence finds its way into policy and practice, as this should have a bearing over how we intervene in these processes. In her review of the theories relating to evidence use in policy, Smith argues that the roots of these different models can be traced back to work done before the 1980s, and that understandings 'do not seem to have progressed significantly over the past three decades, despite numerous new studies' (p. 38). There are many different ideas about how evidence is used in policy and practice and there are tensions between these different theories. Smith points out that public health has tended to operate in a rational and instrumental way, which is at odds with factors such as politics, democracy, ideologies and values. Further, there is little resolution concerning the actual or desirable relationship between research and policy. The mechanisms adopted in most health systems tend to cleave to rather rationalist and instrumental understandings of evidence and practice, where evidence is a good thing that should be used but is undermined often by 'political' factors. Smith argues that instrumental approaches are again reaching their zenith, citing the rise of randomized controlled trials and scientific management approaches (e.g. Haynes et al. 2012). Yet such approaches are only one part of the picture; it is not just a matter of having high-quality research evidence, we also need individuals and organizations that are receptive to particular types of research evidence and have the capabilities and time to apply it in practice. As Smith (2013) also argues based on the UK experience, where multiple research translation institutions exist and do not have cohesion or strategic vision, their impact is limited. This is a salutary lesson for the Australian context.

Internationally, work is underway to provide more politically sensitive, nonlinear and contextualized accounts of research mobilization that draw upon social science theory and which explicate issues of practice (for example, see Swan et al. 2016). The importance of Smith's contribution is to illustrate the many potential understandings of the relationship between research evidence and practice and the fact that these have rarely been resolved within health systems. Further, as Meslin et al. (2013) argue, the focus to date has largely been on the application of research knowledge rather than the processes leading to the creation of effective knowledge and evidence-based guidelines for clinical practice. The journey from idea to creation of research is itself far from simple and involves every degree of politics in terms of how science policy seeks to encourage or constrain innovative research practice. Without understanding the sorts of activities that operate across the whole life course of knowledge translation activities, involving multiple sectors (as in the Triple Helix concept), it is likely that attempts to improve and accelerate these activities will be only partially effective.

Conclusion

Like many other health systems, Australia is currently focused on how it might encourage and accelerate research translation activities for national benefit. In recent years a variety of different institutional mechanisms have been developed to encourage the creation and uptake of high-quality research evidence. This is for the improvement of the health system and citizens, but also in search of the commercial advantage that such mechanisms bring in the face of the changing structure of the Australian economy. In this chapter, we explored three of the mechanisms being used to drive this agenda—Centres of Research Excellence, Advanced Health Research and Translation Centres and Clinical Networks. For each we outlined the nature of these mechanisms and the evidence available locally and internationally. While the evidence base for these organizational arrangements is still emerging, we note an over-all emphasis on governance and structural arrangements, rather than much attention to how these arrangements might develop locally signifi-cant processes of culture change, new capabilities and collaboration in practice. Without a more finely grained focus on how they serve models of research translation, it is difficult to derive lessons for developing such processes more systematically. Further work is needed to understand how these mechanisms operate and interact in the Australian context.

References

Addicott, R., McGivern, G., & Ferlie, E. (2007). The distortion of a mana-gerial technique? The case of clinical networks in UK health care. *British Journal of Management, 18*, 93–107.

Australian Research Council. (2015). *ARC centres of excellence*. Retrieved from http://www.arc.gov.au/arc-centres-excellence. Accessed 14 Mar 2016.

Brooks, P. (2009). The challenge for academic health partnerships. *Medical Journal of Australia, 191*, 26–27.

Brooks, P. (2011). *Explainer: Why Australia needs Advanced Health Research Centres.*

CanNet National Support and Evaluation Service. (2008). *Managed clinical networks: A literature review.* Canberra: Cancer Australia.

Commonwealth of Australia. (2015a). *Boosting high-impact entrepreneurship in Australia—A role for Universities.* Canberra: Office of the Chief Scientist.

Commonwealth of Australia. (2015b). *Reform of the federation: Discussion paper.* Canberra: Commonwealth of Australia.

D'Andreta, D., & Scarbrough, H. (2016). Knowledge mobilization and network ambidexterity in a mandated health care network. In J. Swan, S. Newell, & D. Nicolini (Eds.), *Mobilizing knowledge in health care: Challenges for management and organization.* Oxford: Oxford University Press.

Davies, S. M., Tawfik-Shukor, A., & de Jonge, B. (2010). Structure, govern-ance, and organizational dynamics of university medical centres in the Netherlands. *Academic Medicine, 85*, 1091–1097.

Etzkowitz, H., & Leydesdorff, L. (2000). The dynamics of innovation: From national systems and "mode 2" to a triple helix of university-industry-government relations. *Research Policy, 29,* 109–123.

Ferlie, E., Fitzgerald, L., McGivern, G., Dopson, S., & Bennett, C. (2011). Public policy networks and 'wicked problems': A nascent solution? *Public Administration, 89*(2), 307–324.

Ferlie, E., Fitzgerald, L., McGivern, G., Dopson, S., & Bennett, C. (2013). *Making wicked problems governable? The case of managed networks in health care.* Oxford: OUP.

Fischer, M. D., Ferlie, E., French, C. E., Fulop, N. J., & Wolfe, C. (2013). *The creation and survival of an Academic Health Science Organization: Counter-colonization through a new organizational form?* Said Business School Working Paper. Oxford: University of Oxford.

Fischer, M. D., Ferlie, E., French, C. E., Fulop, N. J., & Wolfe, C. (2015). Affective mobilization in major institutional change: Creating an Academic Health Science Centre. *Academy of Management Perspectives,* 12583.

Fisk, N. M., Steven, L., Wesselingh, J. L., Beilby, N., Glasgow, J., Puddey, I. B., Robinson, B. G., Angus, J. A., & Smoth, P. J. (2011). Academic health science centres in Australia: Let's get competitive. *Medical Journal of Australia, 194,* 59–60.

French, C. E., Ferlie, E., & Fulop, N. J. (2014). The international spread of Academic Health Science Centres: A scoping review and the case of policy transfer to England. *Health Policy, 117,* 382–391.

Government of South Australia. (2007). *Statewide clinical networks: A framework for delivering best practice health care.* Adelaide: Government of South Australia.

Haynes, L., Service, O., Goldacre, B., Torgeson, D. (2012). *Test, learn, adapt: Developing public policy with randomised controlled trials.* London: Cabinet Office Behavioural Insights Team.

Jennings, G. L. R., & Walsh, M. K. (2013). Integrated health research centres for Australia. *Medical Journal of Australia, 199,* 320–321.

Khoury, M. J., Gwinn, M., Yoon, P. W., Dowling, N., Moore, C. A., & Bradley, L. (2007). The continuum of translation research in genomic medicine: how can we accelerate the appropriate integratin of human genome discoverise into health care and disease prevention? *Genetic Medicine, 9*(10), 665–674.

Khoury, M. J., Gwinn, M., & Ioannidis, J. P. (2010). The emergence of translational epidemiology: From scientific discovery to population health impact. *American Journal of Epidemiology, 172*(5), 517–524.

McKeon, S., Alexander, E., Brodaty, H., Ferris, B., Frazer, I., & Little, M. (2013). *Strategic review of health and medical research: Final report.* Canberra: Australian Government Department of Health and Ageing.

Meslin, E, Blasimme, A., & Cambon-Thomsen, A. (2013). Mapping the translational science policy 'valley of death'. *Clinical and Translational Medicine, 2*(1), 14.

National Health and Medical Research Council. (2010). *Discussion paper: Developing Advanced Health Research Centres in Australia.* Canberra: NHMRC.

National Health and Medical Research Council. (2015). *About the NHMRC.* Canberra: NHMRC.

Office of the Chief Economist. (2014). *Australian innovation survey report 2014.* Canberra: Office of the Chief Economist.

Ovseiko, P., Heitmueller, A., Allen, P., Davies, S., Wells, G., Ford, G., et al. (2014). Improving accountability through alignment: The role of academic health science centres and networks in England. *BMC Health Services Research, 14,* 24.

Penfield, T., Baker, M. J., Scobie, R., & Wykes, M. C. (2013). Assessment evaluations, and definitions of research impact: A review. *Research Evaluation, 21,* 21–32.

Perri, P., Goodwin, N., Peck, E., & Freeman, T. (2006). *Managing networks of twenty-first century organizations.* Basingstoke: Palgrave Macmillan.

Pittaway, L., Robertson, M., Munir, K., Denyer, D., & Neely, A. (2004). Networking and innovation: A systematic review of the evidence. *International Journal of Management Reviews, 5,* 137–168.

Rowley, E., Morris, R., Currie, G., & Schneider, J. (2012). Research into practice: Collaboration for Leadership in Applied Health Research and Care (CLAHRC) for Nottinghamshire, Derbyshire, Lincolnshire (NDL). *Implementation Science, 7,* 1–11.

Smith, K. (2013). *Beyond evidence-based policy in public health: The interplay of ideas.* Basingstoke: Palgrave Macmillan.

Swan, J., Scarbrough, H., & Ziebro, M. (2016). Liminal roles as a source of creative agency in management: The case of knowledge-sharing communities. *Human Relations, 69*(3), 781–811.

Tetroe, J. M., Graham, I. D., Foy, R., Robinson, N., Eccles, M. P., Wensing, M., & Ward, J. E. (2008). Health research funding agencies' support and promotion of knowledge translation: An international study. *Milbank Quarterly, 86*(1), 125–155.

Thomas, H. (2003). Clinical networks for Doctors and Managers. *British Medical Journal, 326(7390)*, 655.

Thornicroft, G., Lempp, H., & Tansella, M. (2011). The place of implementation science in the translational medicine continuum. *Psychological Medicine, 41*, 2015–2021.

Part III

Agents, Co-producers and Recipients of Quality Care

Part III

13

Framing a Movement for Improvement: Hospital Managers' Use of Social Movement Ideas in the Implementation of a Patient Safety Framework

Amanda Crompton and Justin Waring

Introduction

It is widely acknowledged that programmatic improvements are difficult to realise in healthcare (Dixon-Woods et al. 2012). It is argued, for instance, that clinicians often resist changes that are imposed upon them, or appear to be motivated by managerial or political interests. Where changes are imposed they may have limited congruity with healthcare professionals, and perhaps most significantly they are seen as challenging professional values, identities and jurisdictions. To overcome these well-recognised problems of implementing and sustaining improvement, leaders of change are encouraged to create the necessary 'receptive context' or 'culture' for improvements to be realised (Pettigrew et al. 1992).

A. Crompton (✉) · J. Waring
Nottingham University Business School, Nottingham, UK
e-mail: Amanda.crompton@nottingham.ac.uk

© The Author(s) 2018
A.M. McDermott et al. (eds.), *Managing Improvement in Healthcare*,
Organizational Behaviour in Health Care,
https://doi.org/10.1007/978-3-319-62235-4_13

Reflecting wider transitions in public sector governance, there has been increased interest in collaborative and participatory improvement methodologies that help create receptive contexts (Dixon-Woods et al. 2012). These encourage frontline clinicians to share unique insights and experiences through the co-design or co-production of 'bottom-up' improvements. This is exemplified by the Institute for Health Improvement's (IHI) Breakthrough Collaborative series, bringing together clinical 'learning communities' that participate in structured improvement activities to develop, implement and share best practice. More recently, it can be seen with the upsurge of interest in methodologies such as experience-based co-design, which offer an alternative to evidence-based top-down reforms (Bate and Robert 2006).

Within this context, there has been sustained interest in social movement strategies to engender 'grassroots' change. In broad terms, social movements are associated with collective action that is (usually) orientated towards changing established social or political institutions (Crossley 2002; Jasper 2010). For healthcare improvement advocates, social movements offer novel lessons for engaging and empowering clinicians to shape the implementation of service improvements (Bate et al. 2004a). The popularity of these ideas can be seen with the 'Million Change Agents' framework (Bate et al. 2004b), the NHS England (2016) 'Health as a Social Movement' programme and the Health Foundation's (2016) 'Q Fellows'.

Notwithstanding the potential for social movements to engender healthcare improvement and influence health policies more broadly, in this chapter, we hope to encourage policymakers, improvement advocates and scholars to reflect upon how social movement ideas are adopted and applied as a method of improvement. More significantly, we encourage advocates to recognise the 'dark sides' of social movements, to consider how appeals to empowerment and improvement associated with collective action can mask more insidious change, and to recognise that grassroots change is not always benign in intent.

Specifically, we suggest that the adoption of social movement ideas by some improvement advocates resembles an instrumental strategy for engaging and empowering clinical communities in relatively prescribed 'positive' change. This may exacerbate the undercurrents of power and ideology inherent to healthcare systems and undermine the collective basis that social movements require. We elaborate this view, looking in particular at the way service leaders use framing strategies to orchestrate movement ideas to engage clinicians in 'grassroots' improvement (Wallace and Schneller 2008).

Learning from Social Movements?

For the purpose of our study, we conceptualise social movements as collective action that is manifest through networks of 'grassroots' activists, motivated by the desire to change prevailing social or political institutions (Crossley 2002). We also recognise that social movements can have conservative goals for maintaining social institutions, and may evolve from emergent local action to become developed 'social movement organisations' with formal leadership and structure. In the healthcare context, social movements have garnered attention because of their potential to challenge and transform institutionalised ways of organising and delivering care (Brown and Zavestoski 2004; Banaszak-Holl et al. 2010).

Of particular interest to our chapter is the way quality improvement advocates have adopted social movement strategies to achieve healthcare improvements. In their review chapter, Bate et al. (2004a) argue that many healthcare workers are engaged in top-down improvement initiatives that involve implementing centrally planned and managed change programmes. However, such initiatives often struggle to realise change because they fail to engage and enthuse frontline staff. As an alternative, social movements may tap into the 'latent potential' for change found across healthcare systems, and secure 'wider and deeper participation in a movement for improvement' (Bate et al. 2004a, p. 64).

Drawing on the work of Bate et al. (2004a, b), our chapter is interested in how framing strategies are used by service leaders to build movements for improvement. Frames are social constructs that, when communicated, influence how actors interpret and make sense of a given situation (Goffman 1974). The analysis of frames and framing is a prominent theme within social movement research, which examines how collective narratives are constructed to shape the meanings and motives of individuals, and in turn align individual action with the aspirations of the collective movement (Benford and Snow 2000; Snow 2004; Oliver and Johnston 2000). In practical terms, Benford and Snow (2000) identify three core framing tasks: 'diagnostic framing', identifying the need for action or the problem; 'prognostic framing', defining the parameters of action; and 'motivational framing', identifying what drives engagement and sustained involvement.

Whilst early social movement research focused on the collective and organic nature of movements, thereby downplaying the role of leadership (Goodwin and Jasper 2014), contemporary research suggests that leaders are central to the formation and mobilisation of movements, especially in framing the need for change, inspiring and motivating diverse stakeholders, and devising strategies for change (Ganz 2013; Morris and Staggenborg 2004). For example, Zald et al. (2005) distinguish between senior leaders who determine the 'priorities' for change and middle-level leaders who identify 'possibilities' for change. Developing a more critical interpretation, however, it is possible to see leaders as imposing particular interests upon local communities rather than representing the interests of grassroots communities. This can be seen in (2015) analysis of the *Action for Happiness* movement, which shows how prominent national figures imposed aspiration for change onto local communities. In such cases, politicisation is far from 'bottom-up' but orchestrated by senior advisors.

The Case Study

Our chapter examines how healthcare leaders sought to 'build the movement' (Director of Nursing) to inform the implementation of improvement techniques. Our chapter is not concerned with the techniques themselves, rather the framing strategies used by leaders to engage and empower frontline clinicians in 'grassroots' improvement activities. The research involved an organisational case study of one NHS hospital's use of social movement approaches to implement a portfolio of quality improvement (QI) interventions. The hospital was identified following a desk-based review of QI projects across the English Midlands, where three hospitals were identified as using social movement approaches.

The selected organisation was a medium-sized District General Hospital with around 500 beds, including medical, surgical, emergency and maternity services. Between 2013 and 2015, the Executive Board tasked senior hospital managers and clinical leaders with devising and implementing a revised QI framework that reflected policy recommendations, best practice and innovations in other sectors. The framework comprised of five elements: (a) a 'Stop the Line' and rapid problem-solving technique to address quality concerns (Sugimori et al. 1977); (b) PDSA cycles to address local improvement challenges (Walley and Gowland 2004); and (c) a new incident reporting system to document safety breaches for the purpose of organisational learning (Barach and Small 2000). These were supported by (d) a leadership development programme, and (e) a broadly conceived culture change programme (Berwick, 1994). Our study focused on the utilisation of social movement ideas as a means of implementing this QI portfolio. The implementation of this new framework was explicitly shaped by managers' conscious adoption of social movement ideas to communicate with and engage staff. In this chapter, we examine the framing strategies used to engage, enrol and empower staff in the change initiative.

Data were collected over twelve months and involved a combination of non-participant observations, semi-structured interviews, focus groups and textual analysis. An initial set of interviews (11) were carried out with senior managers (4), senior medical and nursing leaders (2), leaders of the change initiative (3) and quality and safety managers (2). These considered the development of the QI initiatives and the rationale for using a social movement approach. Observations were undertaken in hospital management offices, team briefings, training events and management meetings, focusing in particular on the interaction between management and frontline staff. Over ninety hours of observation were recorded in field journals. A second phase of data collection involved interviews (23) with 'campaign leaders' located across hospital departments, to understand the further operationalisation of the communication strategy. Finally, three focus groups were undertaken with staff from different areas of the hospital, including two focus groups with nursing representatives (10), one with allied health professionals (4) and one with support and administrative staff (5).

Observation records, verbatim transcripts and selected documents were analysed following an interpretative approach (Corbin and Strauss 2014). This involved coding data to describe the framing strategies of leaders and to understand the effect on the wider workforce. Our analysis looked at the way service leaders, following a social movement approach, constructed and communicated framing strategies as a means of engaging and empowering frontline clinicians in a supposedly 'grassroots' improvement campaign. The analysis was informed by Benford and Snow's (2000) classification of framing tasks, where we look at the ways problems are diagnosed, interventions are promoted as offering solutions and beliefs and values are articulated for securing commitment. Although we present these as distinct activities, in many cases they overlapped, with diagnostic frames juxtaposed or interwoven with prognostic frames. We then look at the responses and reactions of frontline clinicians to these different framing strategies, especially whether they help build a movement.

Building a Movement for Improvement: Managers' Framing Strategies

To introduce the findings, it is useful to describe the broader context of managers' framing activities. As outlined above, senior hospital managers had devised a new Quality Improvement (QI) portfolio in response to external and internal pressures for change. Whilst developing this, managers reflected on the past difficulties of implementing QI methods within the hospital, and actively sought innovative methods to engage staff and support the uptake of change. Senior managers reported appraising various approaches, ultimately deciding to follow a social movement-type approach. This idea reflected some senior managers' broader understanding of social movements and also the growing popularity of social movement methods in health improvement. In following his approach, managers developed a range of engagement and communication activities, including workshops, training, celebrations and pledge campaigns. This included the formation of local action groups (LAGs) to 'spread the word' across the hospital. We examine the type of framing strategies used when engaging staff both directly in a variety of forums and indirectly through communication media.

The Patient Safety 'Problem'

In the early days of formulating their 'campaign' to promote the QI framework, managers' interactions with staff tended to highlight two problems. The first and most prominent of these related to patient safety. This was framed in ways that linked broader external pressures to internal issues. The apparent consequence of this was that managers presented themselves as reacting or responding to the need for change, not as the originators of change. In other words, they distanced themselves from the pressures for change.

Looking more closely at how managers framed the problems of patient safety, three interlinked issues stood out in their communications with staff. The first related to the problems experienced in other hospitals and the idea that patient safety was a system-wide problem. As one manager suggested:

> [we need to] remind staff that we are not immune to the problems faced by the wider health service.

The recently published inquiry into poor quality care and patient deaths at Mid-Staffordshire hospital (Francis 2013) provided a powerful reference for managers when engaging with staff. In training and induction events, for example, managers talked about the risks of 'being another Mid-Staffs'. Managers also made frequent references to the 'headline' findings and recommendations from the inquiry, such as the duty of care that all professionals should have for their patients. In this way, managers seemed to be linking the high-profile experiences of this hospital to the need for change, or rather renewed professionalism, within the everyday practices of frontline clinicians to 'safeguard their patients'.

> We have to do these things, we can't afford to be another Mid-Staffs … good is not good enough. That's our mantra. (Manager of Quality)

The second way managers presented the problem of patient safety was to emphasise broader changes in the policy and regulatory landscape. In management briefings, for example, senior managers explained to departmental managers and clinical leads about the expectations and requirements of agencies such as the Care Quality Commission (CQC), and local care commissioning, professional associations and patient representative groups. A forthcoming visit from the CQC was often highlighted as a major driver for change and precipitating the introduction of the new QI framework. Again, managers seemed to distance themselves from the root cause of the change, and present themselves as a 'buffer' between the demands of external 'inspectors' and the internal changes needed across the hospital.

> So, we know the CQC will be paying us a visit and we need to get our house in order. They will be looking at all our governance arrangements … so we all need to make sure we are on top of our game. (Operations Manager)

In contrast, the third way managers talked about the problem of patient safety was with reference to specific issues or concerns detected in the hospital. These were discussed in general meetings, but more often when engaging with clinicians and leaders from individual departments. For example, when meeting with leaders from the operating theatres reference was made to a recent incident involving missing swabs, and when speaking with doctors and nurses of the elderly care ward reference was made to patient falls. In this way, hospital managers used existing incident reporting and risk management data to link the wider expectations for change to local issues that front-line clinicians could identify with. This not only made the need for change seem more real to clinicians, but also made it difficult for clinicians to offer any opposition; as managers seemed to be targeting documented 'problems' within these areas to justify change:

> We know you've had problems, every department has ... things like patient falls will happen... What you've got to do is make it so they are less likely to happen and when they do happen we all learn. (Presentation to Care of the Elderly Ward)

There was also evidence of subtler diagnostic framing around the problems of implementing change and improvement within the hospital. This was largely overshadowed by the broader problems of patient safety, but it had an important role in justifying the particular 'campaign' approach adopted by managers. A common concern amongst managers was that they felt front-line staff were 'fed-up' with change, and that there was change fatigue across the hospital:

> We know you have had a lot of change to deal with. We've tried several things in the past and not all of them have been as successful as we hoped but that doesn't mean we can stop trying to do things better.

Managers also explained to staff that many of the problems of implementing change in the past were down to the naïveté of senior leaders in thinking change could be imposed upon staff or that structural changes were the only way to change frontline practices. By highlighting their previous shortcomings, managers seemed to be representing themselves, and the approach now being taken, as in some way different or more mature. In several meetings, for example, managers talked about their own learning, which largely centred on recognising that change had to come from the clinicians themselves, and that managers could, at best, support and facilitate the change process. Again, this seemed to de-emphasise the agency and influence of managers, and relocate responsibility for improvement with frontline clinicians:

It has to come from you. I can't make your service safer. Only you know what is going on, and my role is to make it easier for you to make things better.

Learning from Others for Grassroots Improvement

To communicate with staff about the problems facing the hospital, it was common for hospital managers to promote ideas and solutions for how these 'diagnosed' problems might be resolved. This type of prognostic framing focused on the potential for certain interventions to enhance patient safety, but also included more subtle suggestions about how frontline staff might implement these interventions.

Managers' interactions with staff often involved explaining and justifying the proposed QI framework: Stop-the-Line, PDSA, incident reporting, leadership development and culture change. This was framed along three lines. The first was to argue that these solutions were based on QI methods developed and used successfully in other 'high-risk' or 'high-performing' industries. In training sessions, for example, both senior and departmental managers highlighted the 'proven' benefit of the Toyota Production System (or Lean Thinking), how PDSA was internationally recognised as an effective method of problem-solving and how incident reporting was commonplace in the aviation sector. These lines of reasoning are widely accepted and resemble something of a cultural trope within the 'folklore' of QI, articulating unquestioned assumptions about the benefits of 'borrowing' improvement methods from other sectors:

> We have had this [incident reporting] for several years now, but we are far off the likes of BA or Virgin in their safety reporting. It's not just about the serious events, it's the everyday things that we take for granted. (Departmental Manager)

The second framing strategy used to justify the proposed method focused on the way such improvement methods had already been successfully translated and adopted:

Look at the car industry. They have been doing these kinds of improvement works for decades and look at how things have improved. Not just car safety, or airbags, you know, but the way they are made, with fewer and fewer defects. (Quality Manager)

For managers, this demonstrated that these ideas could be effectively integrated into healthcare practice, and that their hospital could replicate the performance improvements witnessed at other exemplar hospitals. Although reference was occasionally made to other regional NHS hospitals, especially a local teaching hospital, managers more often talked of the approaches developed in famous US hospitals, such as the Virginia Mason hospital. One Quality Manager frequently made reference to Charles Kenney's book *Transforming Healthcare* which described the improvement made at Virginia Mason. This text took on some form of sacred status with senior hospital leaders and the Quality Manager distributed copies to department leaders and trainers.

> It's a brilliant book. It shows how hospitals can, or should be run. It's not rocket science or anything, really, but what is impressive is how they have achieved it. (Manager of Quality)

The third justification for adopting the proposed QI framework centred on the recommendations made in recent high-profile patient safety reports and inquiries, especially the Mid-Staffordshire Report (Francis 2013). In particular, managers focused on the need for culture change, so that patient safety, compassion and the sense of duty was central to all aspects of work. Significantly, managers seemed to suggest that the most effective ways to create a safety culture were through embracing the proposed QI interventions, because, as outlined above, they have proven utility in assuring safety.

As well as justifying the proposed QI methods themselves, managers also talked with staff about how these methods could be more effectively implemented through staff taking greater responsibility for QI:

> We have got to become a safer service, where patients can feel confident in the care we give to them. You can do this easily by reporting things

that concern you, by putting your hand up and saying 'stop' when you are concerned, by constantly asking questions of how you can make patient care better. (Trainer)

Although managers rarely talked openly with clinical staff about following a social movement approach, they often talked about the change process as a 'campaign', asked staff to 'pledge' support and routinely made reference to the idea that frontline staff could take ownership of interventions.

> The Trust has thought a lot about how we can work together, we really want to avoid a sense of you and us... We want to help you to help yourselves. And that's what we think the framework will do. (Quality Manager)

Considerable emphasis was placed on providing staff with a broad 'set of tools', but with the espoused expectation for frontline staff to use these tools within the context of pre-existing clinical governance. As such, managers presented themselves as supporting and enabling, rather than commanding or managing staff. Part of this was to encourage staff to participate in LAGs, which championed the proposed QI methods and offered focused training and support for clinical departments. Although these groups appeared to be concerned with supporting clinicians to work with specific QI methods, it was also clear that they offered staff limited scope to modify QI methods or devise alternative techniques. Despite the claim to promote local ownership, in many ways these groups often seemed more concerned with managing the implementation of change, but in ways that gave the impression of local ownership.

> There is a timetable for implementation, and we are working with clinical teams to make sure everything is up and running by the launch date. A lot of what we are doing is training, showing colleagues how to run an effective 'swarm' [rapid improvement circle] and who to call for support. (Group Leader)

Significantly, managers' engagement with frontline staff often emphasised the idea that patient safety was ultimately the responsibility of
every clinician, as part of their professional duty of care. Despite framing patient safety as a 'system' issue, this approach seemed to relocate
responsibility for quality and safety (back) with clinicians, whilst re-
casting managers and service leaders as responsible for ensuring the
necessary QI methods are in place, and staff are appropriately trained;
rather than having direct responsibility for safety.

Reactions at 'Grassroots' Level

There was widespread agreement amongst hospital staff that the quality
and safety of patient care was a priority. Staff also appreciated that external and regulatory factors placed considerable pressure on the hospital
to improve standards and care quality. Significantly, clinicians seemed to
be of the view that hospital managers were not the source of such pressures, and therefore not 'behind' the new QI framework; rather, managers were seen as necessarily responding to these pressures on behalf
of the hospital. As such, managers were, to some degree, successful as
framing themselves in a less strategic and more responsive light, which
might account for clinicians' relatively sympathetic response:

> What with the CQC visit and the demands of commissioners and the
> Department [of Health], it is no wonder the exec are putting in a new
> strategy. (Departmental Manager)

> We don't want to be another Mid-Staffs. It was terrible what happened
> there and it's so easy to forget about the simple things. So anything that
> helps with that is welcome. (Ward Nurse)

Despite broad support for the need for change, some were more critical about the planned changes across the hospital. Although staff were
familiar with incident reporting and PDSA, many were sceptical about
the learning and improvement these tools enabled. As shown by others, doctors were especially critical of the ways these systems were opera-
tionalised (Waring 2005). In particular, doctors were critical of the way

reporting and risk management processes were aligned with managerial processes and decision-making and not with local governance arrangements. Some described how alternative forms of case review and peer review could be equally useful in promoting improvement. Similarly, PDSA was seen by some as 'beguiling simplistic' with the assumption of reviewing performance, but that the reality of undertaking PDSA could be time-consuming and complex. Underpinning these views there appeared to be a deep-seated concern about the use of improvement techniques from other industries:

> PDSA is a lot more complicated than they let on. It's not just a four stage audit process, it requires proper resourcing and specialist skills to manage the process. (Doctor)

More significant criticisms were reserved for the way managers articulated the idea that clinical teams would have significant influence and control over the QI framework. There was widespread support for the idea that staff could tailor and modify interventions to align with pre-existing procedures or local needs, but many questioned whether this was really possible:

> They have told us that we can change how we report locally, but when we asked to change the form and data capture, we were told we couldn't... So I am not sure what we can change. (Ward Manager)

Despite many participants recognising that a lack of clinical engagement had hampered past improvement initiatives, there remained scepticism that the types of engagement described by managers was in any sense 'real'. For some, the campaign approach and the introduction of LAGs to 'spread the word' seemed superficial and contrived. In this sense, some clinicians saw it as an underhand way for managers to influence staff without giving the impression of influence.

> If you look past all the glitz of the campaign and actually look at what is being implemented, it is just another improvement policy, and all this talk of doing things differently seems like a smoke-screen. (Doctor)

What they are proposing is different and I like that, it shows they (Executive) are willing to try new things. But really I am not sure they mean what they are saying about us having local control and us shaping the agenda. (Senior nurse)

Concluding Remarks

Our study examined how hospital managers adopted social movement ideas to promote 'grassroots' quality improvement. Focussing on the framing strategies used by managers, we found interlinked examples of diagnostic, prognostic and motivational framing (Benford and Snow 2000). Diagnostic framing was primarily constructed around the problem of patient safety, which, significantly, presented hospital managers as more passive conduits for reform rather than strategic operators. This might be seen as a strategic framing technique given well-documented instances of professional resistance to more proactive forms of management. It was also significant that managers talked relatively less about the problems of implementing change, focussing on the problem of safety instead of the problem of changing clinician behaviours. This might be because such issues were expected to provoke concern and resistance amongst staff, and it was therefore more prudent to focus on the problem of safety; as one manager said, 'no one can argue against improving patient safety'. Echoing this, managers' prognostic framing centred on proven techniques for improving quality, drawn from other industries or exemplar healthcare providers. Again, there was relatively little emphasis on the type of campaign or movement approach. Where this did become clearer was in relation to frontline clinicians having more influence or control on how the proposed QI methods were implemented and operating locally. Here, LAGs, comprising senior clinical leaders, were presented as supporting staff to work with the new or revised procedures. However, clinicians raised concerns about the extent to which this local influence was real, and saw the changes as often prescribed, which created some tension as it potentially threatened the autonomy of healthcare professionals. To overcome this, managers repeatedly developed more motivational frames around the importance of caring for patients, improving the quality of care and restating the importance of professional duty.

In the processes of building or mobilising a social movement, framing involves constructing particular problems, and the solutions to these problems, in ways that attract and align individual interests to those of the movement, and as a basis for collective action (Benford and Snow 2000; Oliver and Johnston 2000). In our study, the framing centred on the problem of patient safety and the relevance of the proposed QI methods. There was little or no mention of the need for collective or grassroots action, beyond the idea that clinicians should take greater responsibility for patient safety and have scope to influence how hospital policies could be locally implemented. This might suggest that despite growing interest in following social movement-type approaches, and supporting grassroots or emergent change, the managers in our case were not explicit about following this approach.

Earlier, we asked more critical questions about whether leaders 're-align' interests to reflect those of the prescribed movement agenda. Our study found that managers were strategic in the selection of issues and interests to focus on in their framing activities, which positioned them as not forcing change upon staff, and as giving staff greater opportunities to influence change. However, the study also found that managers had a clearly worked out and relatively prescribed QI framework, and that staff had only limited scope to influence the form and operation of this framework. It might be argued that managers' use of a social movement approach, as reflected in the framing activities, was a more deceptive strategy for countering resistance and securing professional support for what, at face value, promised to be emergent and locally owned, but might also be seen as highly prescribed. As such, managers' adoption of social movement ideas in our case study seemed to have little concern with fostering and framing bottom-up improvement work, but function rather as a means for reducing resistance to a relatively prescribed top-down improvement framework.

References

Banaszak-Holl, J., Levitsky, S., & Zald, M. (2010). *Social movements and the transformation of American Health Care*. Oxford: Oxford University Press.

Barach, P., & Small, S. (2000). Reporting and preventing medical mishaps: Lessons from non-medical near miss reporting systems. *British Medical Journal, 320*(7237), 759.

Bate, P., Robert, G., & Bevan, H. (2004a). The next phase of healthcare improvement: what can we learn from social movements? *Quality and Safety in Health Care, 13*(1), 62–66.

Bate, P., Bevan, H., & Robert, G. (2004b). *Towards a million change agents. A review of the social movements literature: Implications for large scale change in the NHS*. Leicester: NHS Modernisation Agency.

Bate, P., & Robert, G. (2006). Experience-based design: From redesigning the system around the patient to co-designing services with the patient. *Quality and Safety in Health Care, 15*(5), 307–310.

Benford, R., & Snow, D. (2000). Framing processes and social movements: An overview and assessment. *Annual Review of Sociology, 26*(1), 611–639.

Brown, P., & Zavestoski, S. (2004). Social movements in health: An introduction. *Sociology of Health & Illness, 26*(6), 679–694.

Corbin, J., & Strauss, A. (2014). *Basics of qualitative research*. London: Sage.

Crossley, N. (2002). *Making sense of social movements*. London: McGraw-Hill.

Dixon-Woods, M., McNicol, S., & Martin, G. (2012). Ten challenges in improving quality in healthcare: Lessons from the Health Foundation's programme evaluations and relevant literature. *BMJ Quality & Safety*. doi:10.1136/bmjqs-2011-000760.

Francis, R. (2013). *Inquiry into the events at mid-Staffordshire NHS Foundation Trust*. London: TSO.

Frawley, A. (2015). Happiness research: A review of critiques. *Sociology Compass, 9*(1), 62–77.

Ganz, M. (2013). Leading change: Leadership, organization, and social movements. In N. Nohria & R. Khurana (Eds.), *Handbook of leadership theory and practice*. Cambridge, MA: Harvard Business Press.

Goffman, E. (1974). *Frame analysis: An essay on the organization of experience*. Cambridge, MA: Harvard University Press.

Goodwin, J., & Jasper, J. (Eds.). (2014). *The social movements reader: Cases and concepts*. Chichester: Wiley.

Health Foundation. (2016). The Q initiative. Retrieved from: http://www.health.org.uk/programmes/the-q-initiative.

Jasper, J. (2010). Social movement theory today: Toward a theory of action? *Sociology Compass, 4*(11), 965–976.

Morris, A., & Staggenborg, S. (2004). Leadership in social movements. In D. Snow, S. Soule, & H. Kriesi (Eds.), *The Blackwell companion to social movements*. London: Wiley.

NHS England. (2016). Health as a social movement. Retried from www.england.nhs.uk/ourwork/new-care-models/vanguards/empowering/social-movement/.

Oliver, P., & Johnston, H. (2000). What a good idea! Ideologies and frames in social movement research. *Mobilization: An International Quarterly, 5*(1), 37–54.

Pettigrew, A., Ferlie, E., & McKee, L. (1992). Shaping strategic change—The case of the NHS in the 1980s. *Public Money & Management, 12*(3), 27–31.

Snow, D. (2004). *Framing processes, ideology, and discursive fields. The Blackwell companion to social movements* (pp. 380–412). Oxford: Blackwell.

Sugimori, Y., Kusunoki, K., Cho, F., & Uchikawa, S. (1977). Toyota production system and kanban system materialization of just-in-time and respect-for-human system. *The International Journal of Production Research, 15*(6), 553–564.

Wallace, M., & Schneller, E. (2008). Orchestrating emergent change: The 'hospitalist movement' in US healthcare. *Public Administration, 86*(3), 761–778.

Walley, P., & Gowland, B. (2004). Completing the circle: From PD to PDSA. *International Journal of Health Care Quality Assurance, 17*(6), 349–358.

Waring, J. (2005). Beyond blame: Cultural barriers to medical incident reporting. *Social Science and Medicine, 60*(9), 1927–1935.

Zald, M. N., Morrill, C., & Rao, H. (2005). The impact of social movements on organizations. In G. F. Davis, D. McAdam, & W. Richard (Eds.), *Social movements and organization theory* (pp. 253–279). Cambridge: Cambridge University Press.

14

Institutional Work and Innovation in the NHS: The Role of Creating and Disrupting

Kath Checkland, Stephen Parkin, Simon Bailey and Damian Hodgson

Introduction

Managing change in public services has become the normal state of affairs. Hartley (2005) identifies 'eras' of governance in public management, and argues that the current focus upon networked modes of governance (Lowndes and Skelcher 1998) embodies and embeds an

K. Checkland (✉)
Centre for Primary Care, University of Manchester, Manchester, UK
e-mail: Katherine.checkland@manchester.ac.uk

S. Parkin
Health Experiences Research Group, Nuffield Department of Primary Care Health Sciences, University of Oxford, Oxford, UK

S. Bailey · D. Hodgson
Alliance Manchester Business School, University of Manchester, Manchester, UK

© The Author(s) 2018
A.M. McDermott et al. (eds.), *Managing Improvement in Healthcare*,
Organizational Behaviour in Health Care,
https://doi.org/10.1007/978-3-319-62235-4_14

237

assumption of continual innovation in ever-changing contexts. The UK NHS is no exception, with the past twenty years characterised by the continual evocation by policymakers and NHS leaders of the need to 'reform' (Secretary of State for Health 2000; Lewis and Gillam 2003; Department of Health 2006). A major structural reorganisation in 2012 (Health and Social Care Act 2012) passed responsibility for management of the service to an 'arms length' body known as 'NHS England' (NHSE). In 2014, the new leader of NHSE, Simon Stevens, issued the 'Five Year Forward View' (NHS England 2014), highlighting the fact that the NHS faces a £22 billion shortfall, and arguing that managing this requires innovation. Solutions offered focus upon breaking down 'barriers' between primary, secondary, community and social care, and encouraging the establishment of new organisations across these boundaries.

In this chapter, we use the lens of institutional work (Lawrence et al. 2011) to explore the implementation of relatively small-scale innovations across organisational boundaries in the NHS. We offer a contribution which addresses an overlooked issue in fields characterised by plural and complex institutional logics: namely, the practical insights seen through the lens of 'institutional work' which have the potential to support local level innovation (Lawrence et al. 2013). This is particularly important as NHS staff are engaged in the rapid creation of new organisations (Bostock 2015). We highlight the need for micro-creation and micro-disruption, and show how (neither creating new institutions, nor disrupting existing ones) our actors were engaged in continual skilled acts of institutional work which nudged organisations closer together, creating accommodations between competing logics.

Institutions, Institutional Logics and Institutional Work

Institutional perspectives offer a variety of concepts for analysing the interaction between organisations and their environments. Their uptake within organisational analysis over the last three decades can broadly be characterised according to three movements: the new institutionalism

(NI), institutional logics and institutional work. Scott (2008) provides an archetype for the NI perspective, arguing that institutions are characterised by three pillars: *rules* (embodying the regulatory framework governing action); *norms* (representing collective understandings about what is appropriate); and *cultural-cognitive assumptions* (representing deeper assumptions about what is/is not possible within a particular institutional context—or *field*). It has been argued that the UK NHS can usefully be thought of as an organisational field, populated by individual organisations delivering healthcare services (Checkland et al. 2012), and a number of authors have explored the issue of institutional change in health systems (c.f. Caronna 2004; Currie and Guah 2007; Macfarlane et al. 2011).

The concept of institutional logics was introduced from within the NI perspective by Friedland and Alford (1991) as part of an attempt to contextualise organisational and individual decision-making. Their focus paid attention to 'supraorganizational patterns of activity' and 'symbolic systems' through which activities are conducted and made meaningful. Noting a set of 'central' institutions in Western capitalism (market, state, democracy, nuclear family, Christianity), they recognised the potentially contradictory nature of these institutions, making 'multiple logics available to individuals and organizations' (p. 232). For those who have subsequently taken up the institutional logics perspective (e.g. Greenwood et al. 2010; Thornton et al. 2012), the attempt has been to show how multiple logics create multiple rationales for action, attempting to specify the 'countervailing and moderating effects on self-interest and rationality' via 'differences in cultural norms, symbols and practices of different institutional orders' (Thornton et al. 2012, p. 4).

Thus, logics bring the possibility of agency to institutional theory, through the notion that 'constellations' of logics exist within organisations and fields to which individuals may respond in a variety of ways (Goodrick and Reay 2011). Nevertheless, this is a limited notion of agency, with the 'logics' acting as a facilitating schemata to generate and maintain institutions across time and space.

The concept of institutional work was introduced by Lawrence et al. (2006) in order to make way for an expanded form of institutionalised agency: 'the concept of institutional work is based on a growing awareness of institutions as products of human action and reaction, motivated by both idiosyncratic personal interests and agendas for institutional change or preservation' (Lawrence et al. 2009, p. 6). Driven by the interest in both processes of change and preservation, Lawrence and colleagues (Lawrence et al. 2006, 2009, 2011, 2013) propose three categories of institutional work; 'creating', 'maintaining' and 'disrupting'.

Although the institutional work perspective recognises the importance of conflicting institutional logics, it emphasises the *work* undertaken to manage these, arguing that successful 'segmentation' (Goodrick and Reay 2011) of logics does not happen automatically, but arises out of local acts of institutional work.

Criticisms of agent-led approaches within institutional theory suggest that the influence of agency is typically overstated. Seo and Creed (2002) argue that such concerns provide an illustration of the 'paradox of embedded agency' (p. 223), focusing on the contradictions between institutional determinism (rigid regulatory and normative systems that structure/constrain action) versus agency (action, innovation and autonomy). The question raised by this paradox is:

> How can actors change institutions if their actions, intentions, and rationality are all conditioned by the very institution they wish to change? (Holm 1995, p. 398)

This very brief summary introduces some concepts which will be explored further in this chapter. First, we provide a brief description of the study, and present some illustrations of conflicting institutional logics at work. We then present empirical data, asking what types of institutional work were required to innovate across organisational boundaries. Third, we address the structure-agency question raised above, suggesting a number of conditions which support innovation by allowing individuals to transgress the structural constraints of their situation (albeit briefly and partially). A final discussion offers practical insights from this work for NHS staff engaged in innovation and change.

Background to the Study

In June 2013, a Local Area Team (LAT) of NHS England invited all General Practitioners within the region to submit project proposals for a pilot programme aimed at extending access to primary care, improving integration and including innovative use of information technology. Six proposals were successful, each located in different Clinical Commissioning Group (CCG) areas. Initial funding made available varied across the six projects (from approximately £250,000 to £1m) and, although initially awarded for a six month period, both time and monies were extended.

Despite diversity in geographic location, design and delivery of the six projects, all were comparable in terms of their core aims. For example, all sites proposed sharing records within primary care; three sites proposed sharing records across community, acute and social care; four proposed extended availability of GP appointments; and all sites aimed to reduce attendance at local Accident and Emergency.

In late 2013, the LAT commissioned the National Institute for Health Research (NIHR) Collaboration for Leadership in Applied Health Research and Care (CLAHRC) Greater Manchester (GM) to conduct a twelve-month mixed methods evaluation of the programme. A *quantitative* outcome evaluation sought to identify impact upon access to local services (especially those within primary and secondary care). A *qualitative* process evaluation sought to explore the implementation of innovative practice across the sites. Here, we focus upon findings from the latter. A description of the methods used is available elsewhere (Hodgson et al. 2015).

Institutional Logics Within the Pilots

The projects involved action across organisational boundaries. For some, this was mainly between separate GP practices, but for most, it also involved interactions between primary, secondary, community and social care. The attempt to change was therefore one of reconciling multiple and often conflicting logics (e.g. professional/managerial, public/private).

Sometimes logics were so embedded within particular settings that they were not necessarily perceived by those concerned. One site attempted to bring consultants out of hospital to support GPs to manage patients more effectively in the community. The aim, from the point of view of those in the community, was to reduce attendances at the hospital, thereby reducing costs. However, from the hospital's 'corporate logic' perspective (which emphasises the need to maintain activity to ensure financial balance), moving activity into the community created a 'void' which could usefully be filled by more (revenue-raising) activity.

> We would build into our consultant's job plan commitment to [local community clinic]. One day a week your Wednesday afternoon clinic is at [pilot site]. That's easy for me to sort in terms of job planning, but if historically they're doing a clinic here I then give them to [pilot site] – that then leaves me with a void, doesn't it? A void of activity happening here. So it's a case of what do I fill that with? ... I could say, well, is there actually some more specialist activity that we could be doing? It's bit of juggling, really. (Hospital specialist)

Thus, although most of those concerned work within the NHS, and as such share many underlying assumptions and values (Checkland et al. 2012), nevertheless it cannot be assumed that they fully understand one another or work towards precisely the same goals. Innovating across these boundaries and bridging between these different logics requires *work*, and in the following sections, we illustrate the types of institutional work we observed.

Institutional Work in the Pilots

As noted above, Lawrence et al. (2006) identify three categories of institutional work: 'creating', 'maintaining' and 'disrupting'. For the purposes of this analysis, which examines individuals attempting to do something different, we focus upon 'creating' and 'disrupting'. These can be subdivided as set out in Table 14.1.

Table 14.1 Types of institutional work by category and sub-category (adapted from Lawrence et al. (2006))

Creating	Disrupting
Advocacy	Disconnecting sanctions
Defining	Disassociating moral foundation
Vesting	Undermining assumptions and beliefs
Constructing identity	
Constructing normative networks	
Changing normative associations	
Mimicry	
Theorising	
Education	

Below, we present a sample of our data which illustrates the operation of examples of each of these types of institutional work. Our aim is to show that innovation across organisational boundaries and logics requires a broad repertoire of institutional work, produced by a range of actors. We focus upon those sub-categories which were either most prominent or which provide useful insights into the work done to support these small-scale (potentially disruptive) pilot projects.

Creating

Work in this category aims to establish new ways of doing things (Lawrence et al. (2006). Table 14.2 summarises a sample of our data relevant to this category, organised according to the role of the person doing the work.

Advocacy refers to the mobilisation of political and regulatory support for particular forms of activity (Lawrence et al. 2006, p. 221). In the above interview extract (with a Senior Manager), the senior team identified local lead GPs, who have been linked together in a network to act as advocates more widely. *Constructing identity* is also important, as it is a means by which new ways of doing things become normalised (Lawrence and Suddaby 2006, p. 221). Our extract illustrates the leader of a particular scheme making a claim about the nature of the programme they have established. This isn't only about providing particular services, or doing things in particular ways, but extends to optimising

Table 14.2 Data extracts illustrating 'creating'

	Project leads	Clinicians	Managers
Creating	*Creating normative networks* I think the other key thing is we don't have this as a stand-alone, it needs to link in with the other pieces of work, the other programmes—strategically what you want to do anyway. We already strategically wanted to develop primary care. As a CCG, we already have a vision for primary care. We knew what we wanted to do with it. Luckily, we already had shared it with our practices, we'd had good engagement. … you don't start with a programme like this thinking, 'another wonderful new idea'. This is all about, how can we develop and deliver the shared vision that we already had? *Creating identity* Because the thing is we have in a sense been given a mandate to do something, and what we want to achieve is, we want to achieve a good quality health and social care system that optimises the health and well-being of our population. That's what we want to do. What we've been handed by having the pilot is a mandate to do things in a different way and to question things, right?	*Changing normative associations* We've had to park a number of issues over the years. As with every sort of primary care organisation, any CCG, there are practices that have been at loggerheads with each other over different things in the past. We've had to leave that aside, we've had to move forward. We've even accepted working this pilot out of somebody else's building, we've done that. So … we've worked together and I think that you can only do that in the short term if you have trust, you can only do that in the long term if you see the benefits.	*Advocacy* As you probably know, most GPs are feeling battled and under pressure and burnt out. So when you get a letter saying, here's more work for you, 90% of them just put it straight in the bin. But the keen ones, people like [GP X], put his hand up, just because he thought the CCGs should have a pilot site. So it was quite a risk for him to do it in the first place, it was fantastic. But we've had a chance to go round the four or five and get them to do a bit more of a network or a bit more of an interesting model

population health. This claim seeks to establish an identity for the pilot that goes beyond the mundane.

The construction of normative networks within institutional settings involves creating new inter-organisational connections, through which practices become normatively sanctioned (Lawrence and Suddaby 2006, p. 221). In our data extract, a Project Lead explains that their pilot was a manifestation of a shared vision that pre-dated the initiation of the project. The new project becomes 'normalised' by its association with the Clinical Commissioning Group's (CCG) predetermined strategic vision.

Creating may also require the alteration of pre-existing normative associations. Here, creating involves remaking connections between practices that may extend to the moral and cultural foundations under-lying those practices (Lawrence et al. 2006, p. 221). This form of creat-ing may require some degree of compromise between agents in which *trust* also plays a significant role. Our extract highlights the work done by clinicians in setting aside previous assumptions and '*even* work-ing out of someone else's buildings' (our emphasis). This illustrates the extent to which things which might from a managerial perspective seem obvious or uncontroversial may, according to a different logic, appear significant or difficult.

Thus, a strongly established norm—the importance of *place* in service delivery—is set aside, and a new set of norms about service delivery is *created*, enabled by developing trust.

Disrupting

Institutional work considered to be 'disrupting' seeks to change the practice, rules and technology associated with a particular logic (Lawrence and Suddaby 2006, p. 235). Examples of each of the various categories of disrupting practice were noted within the pilot sites, with specific examples presented below (see Table 14.3).

Undermining assumptions and beliefs is important in minimising the perceived risks of innovation (Lawrence and Suddaby 2006, p. 235). In our example above, the project lead is disassociating the pilot from the

Table 14.3 Data extracts illustrating 'disrupting'

	Project leads	Clinicians	Managers
Disrupting	*Undermining assumptions and beliefs* The coordinator integrated work, how do we actually measure the success of integrated working? To a lot of people, integrated means you have one bloody organisation and you group a load of tribes on different floors and call it integrated. Well, I actually want to change the way people think and actually work together.	*Disconnecting sanctions* [we say] … we want you to do this, this and this but we want you to be autonomous and make some decisions. That's mixed messages really. So we say, look, do what you feel comfortable with, we'll work it around, we'll trust you. If it goes tits up, we'll say it's our fault, basically.	*Disassociating moral foundations* It's about the cultural behaviour and the shift of that because I think it's so crucial because they're not used to working in this way. And I keep saying that we're like a Marks and Spencer's, really, because you don't go to Marks and Spencer's in the evening to be told, 'oh, actually, we're open out of hours'. We're open from whatever time and that's the ethos that we want to try and adopt.

potentially threatening incorporation of separate groups within a single organisation, and instead emphasising the (less threatening?) need to alter mind sets and working practices.

Another form of 'disrupting' involves the conscious disconnecting of sanctions that may be associated with a particular undesired practice/activity (Lawrence and Suddaby 2006, p. 235). The Project Clinical Lead in our extract describes developmental work within a care home. This person is claiming that the project leaders are prepared to shield local actors from negative consequences should the new service not work out as planned. This 'disconnection of sanctions' seeks to further reduce the perceived risks associated with innovating.

Finally, an actor's conscious disassociation of practice, rules or technology from organisational and moral foundations represents further disruption of institutional norms (Lawrence and Suddaby 2006, p. 235). This act of disassociating essentially liberates individuals from acting within those norms. This category was strongly represented within our data, as those involved sought to explain why existing ways of working (often referred to as 'NHS Culture') needed to change. What is interesting in the extract above is that the new 'moral foundation' offered is associated with a more commercial approach, which would not usually be considered appealing to an NHS audience.

Summary

We have provided illustrations of two of the types of 'institutional work' identified by Lawrence and Suddaby (2006). This represents a small selection of the data we collected. In the full dataset, we identified illustrations of all eighteen of Lawrence et al.'s categories, suggesting that the successful implementation of change across organisational boundaries requires a combination of different kinds of institutional work. In our extracts, we have highlighted the role of those involved, differentiating project leads from clinicians and managers. This is not intended to imply that only such actors demonstrated such types of work. Rather, it represents an attempt to highlight the fact that those steeped in particular logics may have expertise in or be required to perform particular types of work, depending on the institutional boundaries at issue. Thus, for example, our example of a clinician 'creating' by changing normative associations highlights the extent to which an 'insider' can understand and mitigate concerns. In this example, a manager may not have understood that working from another's building might be a problem. It is beyond the scope of this chapter to quantify or determine exactly which types of work are required by whom in which circumstances. Instead, it is our aim to establish the breadth of institutional work being undertaken. In the next section, we aim to address the paradox of embedded agency as noted above.

Creating and Disrupting as an Empirical Demonstration of Embedded Agency?

These brief illustrations demonstrate creative and disruptive work by a wide variety of individuals who may be regarded as exemplars of 'embedded agency' as a result of their professional position within various healthcare organisations. However, the paradox of each and every illustration provided above is that these actions, whilst agency-led, are underpinned and informed by various institutional pressures associated with the NHS. Accordingly, there is perhaps a need to further demonstrate how the various activities illustrated may be regarded as innovative and autonomous action.

Seo and Creed (2002) provide a dialectical framework for explaining the paradox of embedded agency within organisational settings. They argue that contradiction is an inherent by-product of institutions, which suggests that agency (which they term 'praxis') is a latent condition of all institutional settings, which may be given expression through a variety of enabling conditions. These enabling conditions include the introduction of field-level opportunities that may purposely aim to disturb a particular setting. In this case study, the availability of funding from NHS England provided opportunities for individuals and groups to introduce particular projects. This field-level (structurally led) opportunity allowed a form of praxis (individual action) to occur by bringing 'misaligned interests' to light, and making them available to actors already embedded within the setting concerned (Seo and Creed 2002, p. 232).

A second enabling condition that may explain the contradiction of embedded agency relates to the organisational field's *characteristics* and the extent of organisational heterogeneity. In theory, Seo and Creed (2002, p. 228) argue, field-level heterogeneity can bring to light contradictions between intra- and inter-organisational conformity pressures. Thus, in spite of the shared 'core purpose' of the pilots, inter-organisational work required the purposeful crossing of boundaries, resulting in changes which might not have been considered legitimate at the intra-organisational level. Accordingly, the extent of organisational heterogeneity in this study, and the sense of urgency created by the temporariness of the programme, perhaps enabled individual action to transgress its own embedded conditions.

Lastly, we note the more individual characteristics of praxis that might effectively engage institutional contradictions of structure versus agency that pertain to the actions described above: the social position and specific social characteristics of key actors within a particular institution. Seo and Creed (2002, p. 230) describe effective praxis as demanding a combination of 'critical awareness of social conditions' together with the ability to mobilise 'multilateral and collective action to reconstruct the existing social arrangements'. Clearly, such individuals must occupy key positions within organisational fields and have greater access to resources. To this we can add that our evidence suggests that those who bring an understanding of relevant normative assumptions and associations may be best positioned to do the work required to provide the required modifications to support change. Similarly, those actors with professional positions considered legitimate by stakeholders are those more enabled to build influential bridges across and within organisations (Battilana 2011). Furthermore, those individuals considered to be reflexive to organisational concerns and who can demonstrate social skills that influence empathy and co-operation are regarded as further enabling characteristics of embedded agency. These particular attributes suggest that a degree of organisational leadership can be born from opportunity and autonomy within a given setting. Indeed, these attributes of social position, social characteristics and individual leadership are arguably the *drivers* that underlie the collective body of work associated with all of the projects involved in this study. The following extract is one of many in our data emphasising the enabling role of *locally situated* embedded agency within particular structural settings. It is important to note that *formal* organisational position was not the most important characteristic; however, voluntarism and personal drive were vital:

> I think the programme always gives you a slightly false sense of security – we put the project together, we get it approved, we get the funding … and we do more than what we're expected to do because we enjoy it. And it's our baby and we want to see new ways of working, and essentially it's a blank piece of paper where I can create what I want to create. And that always is an incentive to, kind of, go beyond the bar, kind of thing, and then once you prove something and then it goes mainstream, not everybody has the same philosophy, because now it's being imposed on them. And that's where I think part of the hurdle is, of how do you transfer that enthusiasm as well – that it's a good thing to do. (GP Lead)

Indeed, this extract infers a further paradox: once an initiative is deemed to have 'worked' and new ways of working become the 'new norm', the excitement (and sense of agency) which enabled transformation may be lost.

Discussion

In this chapter, we have illustrated myriad micro-level acts of institutional work associated with change programmes which cross organisational boundaries and require bridging between differing institutional logics.

In describing our findings, we have sought to address the enduring paradox of embedded agency, highlighting a number of conditions which may have supported individuals and groups as they sought to act outside the institutional structures within which they work. This leads on to the important question of how far such evidence can be useful in a practical sense to those currently engaged in an extensive programme of change in the NHS (NHS England 2014). It is clear that, notwithstanding our categorisations, the work being done in our sites was to a large extent instinctive and practical in focus. Our participants were not *trying* to create new normative networks; they were striving to solve practical problems, and the work that they did was the work which appeared to them to be necessary.

One approach to generating wider learning from studies such as this, which focus upon particular change programmes, is to highlight enabling or inhibiting conditions. Thus, for example, Best et al. (2012) synthesised findings from studies of large-scale organisational change in health systems and identified enabling factors such as 'attend to history', 'engage physicians' and 'establish feedback loops'. Our study also generated such conditions, and the study report pointed to such factors as pre-existing networked working and issues associated with information technology and governance as being important (Hodgson et al. 2015). However, using the theoretical perspective of institutionalism and institutional work has perhaps pointed to some more fundamental issues which might usefully be considered by those embarking on change programmes which cross organisational boundaries and which require accommodations between different logics.

First, our study suggests that those who understand the contrasting norms and assumptions embedded within a field may be best situated to support change. This, in turn, implies a need to spend time diagnosing such issues at an early stage. A collective consideration of the world views represented by those who must work together may be possible and useful; such approaches have been shown to be helpful in other contexts (Checkland and Scholes 1989).

Second, our study has highlighted the social position and personal characteristics of the local 'entrepreneurs' who sought to drive change. Other studies have also highlighted the importance of such individuals (Best et al. 2012); our contribution is perhaps to uncover some of the issues which such individuals might be encouraged to focus upon, with a conscious consideration of whether and when creation or disruption might be required, and what types of work might be appropriate. Most research in this tradition, including ours, focuses upon retrospective analysis types of work, in a situation in which the endpoint or outcome is known. Future work could explore the prospective use of institutional work by individuals in these roles.

Finally, it is important to note that 'effective disruption' did not necessarily lead to effective policy implementation. Once disruption has occurred, the direction in which things move cannot necessarily be controlled; without embedding change, existing norms may reassert or change may move off in an undesired direction. Thus, new policy directions cannot be initiated without disruption, but disrupting alone doesn't necessarily lead to desired outcomes.

In our study, the contradictory logics which gave rise to institutional work tended to arise across long-standing, macro-level institutional boundaries: primary and secondary care; primary and community care; and medical and social care. These boundaries are underpinned by deeper philosophical differences about the nature of care. The *work* that we observed across these boundaries was, by contrast, local, micro-level and situated in specific contexts. We would contend that some of the creative work we saw was intended to institutionalise new ways of doing things, but the question remains as to whether or not these new approaches will become normalised at the local level, and whether they will have a longer term impact on the macro-level institutions within which they are situated. Thus, if the Five-year Forward View (NHS England 2014) successfully supports the development of a multitude of new organisations across these long-established boundaries, will the multiple micro-level instances of institutional work and change (as documented here) generate wider macro-institutional change? This question is perhaps one which future research on new organisations currently in development in England could usefully address.

Acknowledgements This study is based on research carried out for a project funded by the National Institute for Health Research Collaboration in Applied Health Research and Care (NIHR CLAHRC), Greater Manchester and by NHS England (Greater Manchester). The views expressed in this chapter are those of the authors and not necessarily those of the NHS, the NIHR or the Department of Health. We are grateful to the participants who gave freely of their time, and to the wider project team.

References

Battilana, J. (2011). The enabling role of social position in diverging from the institutional status quo: Evidence from the UK national health service. *Organization Science, 22*(4), 817–834.

Best, A., Greenhalgh, T., Lewis, S., Saul, J. E., Carroll, S., & Bitz, J. (2012). Large-system transformation in health care: A realist review. *Millbank Quarterly, 90*(3), 421–456.

Bostock, N. (2015). *NHS England reveals £200m 'vanguard' areas trialling NHS integration.* GP Online. Retrieved from http://www.gponline.com/nhs-england-reveals-200m-vanguard-areas-trialling-nhs-integration/article/1337555. Accessed 30 May 2017.

Caronna, C. A. (2004). The misalignment of institutional 'pillars': Consequences for the U.S. health care field. *Journal of Health and Social Behavior, 45,* 45–58.

Checkland, K., Harrison, S., et al. (2012). Commissioning in the English national health service: What's the problem? *Journal of Social Policy, 41*(03), 533–550.

Checkland, P. B., & Scholes, J. (1989). *Soft systems methodology in action.* Chichester: Wiley.

Currie, W. L., & Guah, M. W. (2007). Conflicting institutional logics: A national programme for IT in the organisational field of healthcare. *Journal of Information Technology, 22*(3), 235–247.

Department of Health. (2006). *Health reform in England: Update and next steps.* London: The Stationery Office.

Friedland, R., & Alford, R. (1991). Bringing society back in: Symbols, practices and institutional contradictions. In W. W. Powell & P. J. Dimaggio (Eds.), *The new institutionalism in organizational analysis* (pp. 232–263). Chicago: University of Chicago Press.

Goodrick, E., & Reay, T. (2011). Constellations of institutional logics: Changes in the professional work of pharmacists. *Work and Occupations, 38*(3), 372–416.

Greenwood, R., Díaz, A. M., Li, S. X., & Lorente, J. C. (2010). The multiplicity of institutional logics and the heterogeneity of organizational responses. *Organization Science, 21,* 521–539.

Hartley, J. (2005). Innovation in governance and public services: Past and present. *Public Money & Management, 25*(1), 27–34.

Health and Social Care Act. (2012). Department of Health.

Hodgson, D., Anselmi, L., et al. (2015). *NHS Greater Manchester Primary Care demonstrator evaluation*. Greater Manchester: Collaboration for Leadership in Applied Health and Care Research.

Holm, P. (1995). The dynamics of institutionalization: Transformation processes in Norwegian fisheries. *Administrative Science Quarterly, 40*, 398–422.

Lawrence, T., Suddaby, R., et al. (2011). Institutional work: Refocusing institutional studies of organization. *Journal of Management Inquiry, 20*(1), 52–58.

Lawrence, T. B., Leca, B., et al. (2013). Institutional work: Current research, new directions and overlooked issues. *Organization Studies, 34*(8), 1023–1033.

Lawrence, T. B., Suddaby, R., et al. (2006). Institutions and institutional work. In R. Clegg, C. Hardy, T. B. Lawrence, & W. R. Nord (Eds.), *Handbook of organizational studies*. London: Sage.

Lawrence, T. B., Suddaby, R., et al. (Eds.). (2009). *Institutional work: Actors and agency in institutional studies of organizations*. Cambridge: Cambridge University Press.

Lewis, R., & Gillam, S. (2003). Back to the market: Yet more reform of the National Health Service. *International Journal of Health Services, 33*(1), 77–84.

Lowndes, V., & Skelcher, C. (1998). The dynamics of multi-organizational partnerships: An analysis of changing modes of governance. *Public Administration, 76*(2), 313–333.

Macfarlane, F., Exworthy, M., et al. (2011). Plus ça change, plus c'est la même chose: senior NHS managers' narratives of restructuring. *Sociology of Health & Illness, 33*(6), 914–929.

NHS England. (2014). *Five year forward view*. London: NHS England.

Scott, W. R. (2008). *Insititutions and organizations: Ideas and interests*. Thousand Oaks, CA: Sage.

Secretary of State for Health. (2000). *The NHS plan: A plan for investment, a plan for reform*. London: The Stationery Office.

Seo, M.-G., & Creed, W. E. D. (2002). Institutional contradictions, praxis, and institutional change: A dialectical perspective. *The Academy of Management Review, 27*(2), 222–247.

Thornton, P. H., Ocasio, W., et al. (2012). *The institutional logics perspective: A new approach to culture, structure and process*. New York: Oxford University Press.

15

Attaining Improvement Without Sustaining It? The Evolution of Facilitation in a Healthcare Knowledge Mobilisation Initiative

Roman Kislov, John Humphreys and Gill Harvey

Introduction

How do service improvement techniques evolve over time? This chapter focuses on the temporal dynamics and microprocesses involved in

A reworked and expanded version of this chapter was published by Taylor & Francis Group as: Kislov, R., Humphreys, J., & Harvey, G. (2017) How do managerial techniques evolve over time? The distortion of "facilitation" in healthcare service improvement. *Public Management Review*, *19*(8), 1165–1183, doi:10.1080/14719037.2016.1266022.

R. Kislov (✉)
Alliance Manchester Business School, University of Manchester, Manchester, UK
e-mail: Roman.kislov@manchester.ac.uk

J. Humphreys
NIHR CLAHRC Greater Manchester, Salford Royal NHS Foundation Trust, Salford, UK

G. Harvey
Adelaide Nursing School, The University of Adelaide, Adelaide, Australia

© The Author(s) 2018
A.M. McDermott et al. (eds.), *Managing Improvement in Healthcare*,
Organizational Behaviour in Health Care,
https://doi.org/10.1007/978-3-319-62235-4_15

the evolution of *facilitation*, a service improvement approach that can be defined as a concerted, social process of enabling the mobilisation of evidence-based knowledge into clinical practice (Harvey et al. 2002; Berta et al. 2015). Drawing on a qualitative longitudinal case study of a UK-based knowledge mobilisation programme, we describe the following three parallel and overlapping microprocesses underpinning the gradual distortion of facilitation over time: (1) prioritisation of (measurable) outcomes over the (interactive) process; (2) reduction of (multi-professional) team engagement; and (3) erosion of the facilitator role.

Our theoretical analysis highlights the malleability of the 'core' components of managerial techniques compared to product-based innovations and the marginalisation of the sustainability goals of service improvement under the influence of powerful institutional forces. We reveal potential unintended consequences stemming from the adaptation of service improvement approaches to local contexts advocated by the improvement and implementation literature (Bosch et al. 2007; Fervers et al. 2006; Kirsh et al. 2008; Krein et al. 2010; Ruhe et al. 2009). More specifically, we argue that an uncritical and uncontrolled adaptation of facilitation may lead to its distortion, undermining its promise to positively affect organisational learning processes and masking the unsustainable nature of the resulting improvement outcomes captured by conventional performance measurement.

This chapter is organised as follows. The next section discusses the evolution of managerial techniques and outlines a number of context-mediated tensions shaping their practical implementation. The third section introduces facilitation as a service improvement technique and explores its dynamic relationship with the organisational and institutional context of healthcare. The empirical setting and the procedures for data collection and analysis are described in the *Case and Method* section. The *Findings and Discussion* section describes the three micro-processes underpinning the evolution of facilitation and outlines the theoretical contribution of the study and its practical implications. This is followed by a brief conclusion summarising the key messages of the study.

Evolution of Managerial Techniques

In recent decades, the public sector has experienced an upsurge of managerial strategies, tools and techniques aiming to increase its effectiveness and efficiency (Boaden et al. 2008). In addition to their potential positive impact on performance, these managerial approaches are often viewed as rational, modern and progressive, thus enhancing the legitimacy of the adopting organisation (Lozeau et al. 2002). Whilst managerial techniques are often enthusiastically embraced by managers and practitioners involved in service improvement, there is a growing body of critical research highlighting the difficulties and unintended consequences of their practical application in healthcare organisations (McCann et al. 2015; Dixon-Woods et al. 2012; Radnor et al. 2012; Bate and Robert 2002; Powell and Davies 2012).

The fundamental issue is the potential 'compatibility gap' between a set of assumptions underlying the design of a managerial intervention on the one hand, and the actual cultural, structural and political characteristics of the adopting system on the other (Lozeau et al. 2002). This gap can result in a mismatch between the intended and actual use of managerial innovation that has been referred to in the literature as 'the lack of innovation fidelity' (Lewis and Seibold 1993) or 'misalignment between the deliberate and emergent strategic practices' (Omidvar and Kislov 2016; Mintzberg and Waters 1985). According to Lozeau et al. (2002), such a mismatch can take several forms:

- *Customisation* which involves both adapting the technique *and* adjusting the organisational processes;
- *Loose-coupling* whereby the technique gets adopted only superficially, in a ritualistic way, with the functioning of the organisation remaining largely unaffected; or
- *Co-optation,* or *corruption* whereby the technique becomes captured and distorted to reinforce existing roles and power structures within an organisation.

Finally, Lozeau and colleagues use the term '*transformation*' to denote those (supposedly quite rare) cases of handling the compatibility gap where the adopting organisation modifies its functioning to fit the assumptions behind the technique and where, as a result, the actual use of a managerial technique does not significantly differ from its intended use.

Difficulties of translating managerial approaches into healthcare should be analysed in the context of the following three sets of tensions. First, most managerial innovations are likely to have fluid and negotiable boundaries, and can be viewed as a combination of a '*hard core*', which is relatively fixed and stable regardless of the context, and a '*soft periphery*', related to the multiple ways of local implementation (Denis et al. 2002). The greater the uncertainty about the latter, the more scope there will be for customisation, loose-coupling and corruption.

Second, not only is there a mutual influence between the managerial technique and the adopting system (Denis et al. 2002), but the latter is represented by a number of (often conflicting) professional and managerial groups operating at different levels of the hierarchy. For example, professionals have been shown to actively co-opt managerial approaches and internalise them in their practices, thus reversing managerial control (Kitchener 2000; Kamoche et al. 2014). It has also been noted that securing the support of one professional group can lead to the alienation of others (Dixon-Woods et al. 2012; Powell and Davies 2012).

Finally, the implementation of managerial innovation within an organisation is likely to be shaped by the inconsistent policy context. For example, many service improvement techniques display a contradiction between the rhetoric of professional empowerment and the command-and-control procedures for auditing the performance data representing the managerial agenda (Lozeau et al. 2002; Causer and Exworthy 1998; Harrison and Pollitt 1994). Furthermore, managerial techniques can be 'distorted' in a top-down way to fit the policy imperatives for centralisation and target-driven performance management even prior to the 'bottom-up' customisation in the process of local implementation (Addicott et al. 2007).

Facilitation in the Service Improvement Context

Facilitation is a concerted, social process of enabling the mobilisation of (evidence-based) knowledge into (professional) practice (Berta et al. 2015; Harvey et al. 2002). It deploys a specifically designated role ('facilitator') encouraging others to reflect upon their current practices in order to identify gaps in performance, introduce sustainable evidence-informed practice change, enable knowledge sharing within and across organisations and thus improve the outcomes of service provision. Facilitation is usually goal-oriented, follows a team-based approach and incorporates aspects of project management, leadership, relationship building and communication (Stetler et al. 2006; Berta et al. 2015; Kitson et al. 2008; Harvey and Kitson 2015; Kelly et al. 2002; Dogherty et al. 2010). To position the debates around the processes and outcomes of facilitation in the context of the broader literature on the evolution of managerial techniques, it will be useful to summarise these debates along the three generic sets of tensions introduced in the previous section.

Variability of Interpretations

Whilst the role of a facilitator, the involvement of teams, the articulation of performance-oriented goals and the enabling nature of the facilitation process can be viewed as the 'core' of facilitation as a managerial technique, its 'periphery' is relatively wide. Crucially, facilitation is a multifaceted intervention, with facilitators often deploying a variety of other service improvement tools and techniques as appropriate in a given context (Harvey et al. 2002; Baskerville et al. 2012; Dogherty et al. 2010). The facilitator role can be filled by clinical professionals, researchers or managers (Harvey and Kitson 2015). The performance goals of facilitation projects can be specified in a top-down fashion by senior managers, clinicians and researchers external to the improvement teams, or can be collectively determined by the teams themselves (Harvey et al. 2012; Seers et al. 2012). It can be designed as a pre-planned and monitored sequence of stages (Dogherty et al. 2010) or remain deliberately fluid from the outset, allowing for greater flexibility and emergence (Tierney et al. 2014).

Complexity of the Adopting System

Facilitation as a managerial innovation has a number of wide-ranging implications for the adopting organisation: premised on the team-based approach, it cannot be adopted individually, mandating a reconfiguration of routines and responsibilities at the individual, team and organisational levels (Kislov et al. 2014). Involvement in the facilitated improvement projects may differentially affect various professional groups, increasing workload for some of them (Powell and Davies 2012; Tierney et al. 2014). In fragmented contexts, such as the primary care sector, facilitating improvement across multiple organisations can be more problematic than working with intra-organisational project teams with a history of pre-existing relationships (Kislov et al. 2012). Having a facilitated improvement intervention endorsed at a senior level may increase its formal adoption but does not guarantee motivation and enthusiasm of the local improvement teams (Tierney et al. 2014).

Underlying Policy-Level Contradictions

Like most quality improvement approaches, facilitation involves a tension between its formalised, managerialist, goal-oriented aspects, which in the UK context often means aligning improvement work with nationally mandated policy-driven targets (Tierney et al. 2014; Kislov et al. 2016), and the situated, practice-based, emergent nature of team-level learning processes that are seen as fundamental for achieving sustainable change (Kislov et al. 2011; Currie 2007; Berta et al. 2015). On the one hand, as an approach respecting the collective, situated and practice-based nature of learning, facilitation may well achieve the improvement goals accepted by the adopting system; on the other hand, the target-oriented culture can adversely affect the horizontal processes of learning and knowledge sharing that facilitation relies upon (Kislov et al. 2016; Currie and Suhomlinova 2006; Addicott et al. 2006).

To sum up, we view facilitation as a managerial technique aiming to achieve improvement goals through capitalising on the social nature of the organic learning and knowledge sharing processes within and across organisations. Its evolution is likely to be shaped by the negotiability of its 'soft periphery', the multilevel context of its implementation and the tensions played out at the policy level. Exploring these issues can be beneficial for two reasons. First, it can further develop our theoretical understanding of facilitation by shifting the focus of enquiry from facilitators' roles, characteristics and practices as ingredients of prescriptive contextual change (Petrova et al. 2010; Baskerville et al. 2012; Dogherty et al. 2010) towards the influence contextual tensions can exert on the emergent, contestable and often ambiguous process of facilitation.

In addition, such exploration is beneficial for theorising the evolution of managerial techniques in general. Whilst different scenarios for handling the 'compatibility gap' between managerial interventions and the adopting systems have been described, their analyses predominantly relied on multiple case studies, therefore tending to focus on cross-case variability, underlying contextual differences and the resulting outcomes of the evolution (Addicott et al. 2007; Lozeau et al. 2002). The emergent micro-level processes involved in the evolution of managerial techniques and their responses to the interplay between different sets of tensions remain underexplored.

This study addresses these gaps by presenting an in-depth qualitative longitudinal case study of a UK-based knowledge mobilisation programme relying on facilitation to achieve sustainable evidence-informed change in clinical practice across multiple healthcare organisations. By analysing the evolution of the initiative's approach to facilitation over a five-year period, it will answer the following research questions. *How does the interpretation and application of facilitation as a managerial technique evolve over time? How is its evolution shaped by multiple contextual tensions? What implications does this evolution have for achieving sustainable improvement in practice?*

Case and Method

The longitudinal single case study was conducted in a five-year collaborative programme ('Programme') involving a university, a National Health Service (NHS) hospital and primary care organisations aiming to increase the identification of Chronic Kidney Disease (CKD) and improve the management of blood pressure in CKD patients by facilitating the mobilisation of existing health research in day-to-day clinical practice.

The remit of facilitators in the Programme included guiding and supporting multiprofessional *improvement teams* (comprised of a practice manager, a general practitioner (GP) and a practice nurse) that were created in primary care organisations ('general practices'), where research evidence on CKD was to be mobilised. Each year a new group of general practices was recruited, with three phases of the Programme included in the current analysis (Table 15.1). Facilitators were supported in their frontline activities by a *programme team* comprised of a university-based social scientist with an expertise in service improvement, a hospital-based nephrologist and several managers.

A purposive sampling strategy was used, with forty research participants drawn both from the Programme team and participating general practices. Forty-five semi-structured interviews (30–95 min in duration) served as the main method of data collection and were conducted (face-to-face or by phone) in three rounds (2010–2011, 2012–2013 and 2013–2014) to enable longitudinal analysis. Five of the research participants were interviewed twice due to their involvement in the Programme throughout several phases. Two of the co-authors were members of the Programme team in 2009–2012 (the second author as a facilitator and the third author as a social scientist), which enabled them to conduct participant observation of meetings and reflect on their first-hand experiences of facilitation in the first half of the Programme's lifetime.

The interviews were digitally recorded and transcribed verbatim; transcripts were coded and analysed with the aid of NVivo software. The first stage of data analysis was predominantly inductive, involving a series of emergent descriptive codes and following a narrative analytical strategy that aimed at the construction of a detailed story from the

Table 15.1 The evolution of the Programme over time

	First phase (2009–2010)	Second phase (2011–2012)	Third phase (2012–2014)
Composition of the Programme team delivering the facilitated intervention	Non-clinical facilitators (2) Manager Data analyst Nephrologist Social scientist	Non-clinical facilitator Clinical facilitators (2) Managers (2) Data analyst Nephrologist Social scientist	Non-clinical facilitators (2) Clinical facilitators (3) Managers (3) Data analyst (all staff working part-time)
Representatives of the general practices involved	Improvement teams (a GP, a practice nurse and a practice manager in the majority of practices) driving the project in their practices and involving the rest of the practice staff as appropriate	Improvement teams with variable degrees of involvement for different team members; often one member of a general practice driving the project locally; other practice staff often remaining uninvolved	Practice nurses, with other clinical and non-clinical practice staff usually remaining uninvolved
Tools, techniques and events used by the facilitators	Quality improvement collaborative methodology Plan-Do-Study-Act cycles Context assessment Regular meetings with the improvement teams based in the participating general practices Full-day workshops (3) Project close workshop	Elements of quality improvement collaborative methodology Regular meetings with teams/individuals championing the project in their practices Electronic audit tool Half-day workshops (2) Project close workshop Teleconferences (5)	Electronic audit tool Half-day workshops (2) Individual feedback sessions with practice nurses

(continued)

Table 15.1 (continued)

	First phase (2009–2010)	Second phase (2011–2012)	Third phase (2012–2014)
The role of the facilitators	Enabling the general practice staff to implement evidence-informed improvement Facilitating teamwork within general practices and knowledge sharing between them Data collection and monitoring project progress	Enabling the general practice staff to implement evidence-informed improvement Supporting the general practices in using the auditing tool Educating the clinical staff about the management of CKD (clinical facilitators) Data collection and monitoring progress (non-clinical facilitator)	Project management, data collection and analysis, reporting the outcome data to the CCGs (non-clinical facilitators) Auditing general practice registers using the electronic tool (clinical facilitators) Educating the clinical staff about the management of CKD (clinical facilitators)

raw data (Langley 1999). The second stage of analysis aggregated previously identified contextual factors (e.g. emphasis on targets, recruitment patterns, funding etc.) with a number of emerging categories informed by the theoretical framework (e.g. 'team engagement', 'soft periphery', 'facilitation' etc.). Matrix analysis (Nadin and Cassell 2004) was deployed to compare and contrast the three phases of the Programme. Finally, in an iterative process of refining categories, detecting patterns and developing explanations, existing codes and categories were transformed into three main themes [prioritisation of the (measurable) outcomes over the (interactive) process; reduction of (multiprofessional) team engagement; erosion of the designated facilitator role], which reflected the microprocesses involved in the evolution of facilitation as a managerial technique.

Findings and Discussion

We have identified three interrelated and overlapping microprocesses that underpin the evolution of facilitation as a managerial technique in the contemporary context of the English primary healthcare sector (Table 15.2). The first process, *prioritisation of the outcomes over the process*, denotes the gradual loss of interactive elements of facilitation whilst retaining the focus on those tools that provide a quicker and less resource-intensive way to achieve the measurable outcomes of an improvement project. The second process, *reduction of the team engagement*, describes a gradual disintegration and disengagement of multiprofessional teams whose input is crucial for sustaining improvement within their organisations. The third process, *erosion of the designated facilitator role*, captures a major shift from the 'enabling' function, which forms the core of facilitation (Harvey and Kitson 2015; Harvey et al. 2002; Petrova et al. 2010; Stetler et al. 2006), towards more conventional project management, service improvement or clinical activities. An analysis of these strategies provides a number of theoretical insights.

First, these findings enhance our understanding of the interplay between the 'hard core' and 'soft periphery' (Denis et al. 2002) when

Table 15.2 Three microprocesses underpinning the evolution of facilitation as a managerial technique

	First phase (2009–2010)	Second phase (2011–2012)	Third phase (2012–2014)
Prioritisation of the outcomes over the process	"The knowledge transfer is the day to day contact, and the conduit of the knowledge between the project team and the practices that we work with. So, for example, I will conduct site visits to go and visit the teams in the practice and facilitate their improvement. Basically my work is around that..." (Non-clinical facilitator)	"...Having a theoretical model is useful. It's sort of a gold standard you can work towards. But in reality and in practice that may not be possible, so you may have to adapt according to your specific projects or your specific disease area or your specific GP patch." (Non-clinical facilitator)	"...I think there was more about the [audit] tool ... than there was about the actual basics of CKD education." (Clinical facilitator)
Reduction of the team engagement	"...In phase one there was a big emphasis on making sure the whole practice was involved..." (Manager)	"If I was ever involved in a project like this again, it's one of the things that I would really stick my neck out on is that the rest of the practice wasn't involved early enough." (Practice nurse)	"...[GPs promised:] 'We'll do everything to help' and all the rest of it, but in reality they didn't give the nurses any time to do [the project]." (Clinical facilitator)
Erosion of the designated facilitator role	"...[The facilitator] visiting regularly ... for us it didn't feel like a pressure, it was more of a motivation; it helped us keep our enthusiasm." (General practitioner)	"...[The non-clinical facilitator] stepped up then and was doing more of the liaising with stakeholders and recruiting more practices, more office-based. I think he took on more of a management lead ... I've kept him in the loop with what's happening at the practices but ... he's been pretty hands-off going into the practices." (Clinical facilitator)	"...Phase three ... I think that's where we suddenly started doing work for [the general practices] rather than guiding through the work." (Manager)

applied to managerial techniques. Our case study shows that out of the four core elements of facilitation (the facilitator role, the involvement of teams, the articulation of performance-oriented goals and the enabling nature of the facilitation process), only one (the goals) remained unchanged throughout the Programme, with the other elements being fundamentally transformed or even replaced by the tools that were initially seen as secondary elements of facilitation. These developments suggest that managerial techniques have a wider and a 'softer' core compared to product-based innovations, potentially putting them at a higher risk of corruption by privileging some of the core elements at the expense of others *and* replacing these lost core elements by the components of the 'soft periphery'.

Second, we contribute to a theoretical understanding of how managerial techniques evolve over time. Whilst Lozeau and colleagues (2002) present customisation, loose-coupling and transformation as distinct scenarios of closing the compatibility gap between the proposed intervention and the real-world context of its implementation, our data indicate that these could be viewed as temporal stages of the broader evolutionary process. In our case study, the facilitated intervention progressed from transformation in the first phase of the Programme through customisation and loose-coupling at the second phase (when the initial model started to be adapted but its core components were not yet completely lost), to corruption in the third phase, whereby the distorted facilitation approach failed to modify existing roles and power structures in the general practices and was co-opted for producing outcomes prioritised by the most powerful stakeholders. We show that the evolution of managerial techniques is a gradual process, which can be hidden behind the rhetoric pertaining to the initial intervention. For instance, the use of such terms as 'facilitation', 'facilitators' and 'improvement teams' lasted well into the corruption stage, when these terms no longer adequately conveyed the essence of the transformed intervention.

Finally, in contrast to an instrumental view of service improvement as an active reconfiguration of contextual processes and structures (Kitson et al. 1998, 2008; Damschroder et al. 2009), our study highlights the crucial role of organisational and institutional factors in the corruption of service improvement techniques. The eroding effect of policy-driven targets on sustainable service improvement is realised through the duality of goals behind the introduction of managerial techniques. On the one hand, proponents of service improvement approaches recognise the importance of achieving sustainable change through promoting education, collaboration and knowledge sharing (Dixon-Woods et al. 2012). On the other hand, this aspect usually remains implicit, unarticulated and elusive, with the outcomes of the 'sustainability objective' proving difficult to capture in a transparent and quantifiable way, which leads to the prominence of more tangible, quantifiable, target-related improvement goals that are favoured by the current policy context.

This study also raises a number of practical implications for service improvement in general and facilitation in particular. Whilst the need to adapt interventions to local context is widely acknowledged in the health services research literature (Bosch et al. 2007; Fervers et al. 2006; Kirsh et al. 2008; Krein et al. 2010; Ruhe et al. 2009), our data show that whilst in theory facilitators have a wide range of tools and techniques open to them in order to enable evidence-based improvement, in reality the approach that they take is severely constrained by the context that they work in. These constraints trigger the process of customisation and, unless actively counterbalanced, may lead to the corruption of the initial approach, with its sustainability goals remaining unrealised in practice. Education, team engagement and provision of protected time for improvement can be powerful counterbalancing factors but require adequate resourcing (Krein et al. 2010; Kislov et al. 2012; Dixon-Woods et al. 2012).

Another important implication relates to the recruitment and development of designated service improvement roles. Non-clinical facilitators working in healthcare, whilst having strong interpersonal skills and/or improvement expertise (Harvey and Kitson 2015), may lack legitimacy, which can be rectified by the deployment of clinicians to fill these designated roles (Petrova et al. 2010; Shipman et al. 2003). Our findings suggest that whilst both groups of facilitators have strengths, the evolution of these roles in practice demonstrates a common tendency to shift away from the interactive, enabling, facilitative aspects of service improvement. However, this shift takes a different form in each of the two cases.

Non-clinical facilitators switch from *enabling* (frontline facilitation of service improvement) to *managing* (i.e. office-based project management and performance measurement). Clinical facilitators, in turn, demonstrate the shift from *enabling* to *doing* (i.e. involvement in the actual improvement work instead of supporting and educating others), which is made possible by their professional knowledge and skills (Kislov et al. 2016).

Conclusion

The analytical contribution of this study is threefold. First, drawing on the distinction between the 'hard core' and 'soft periphery' of innovation (Denis et al. 2002), we show that the corruption of managerial techniques involves privileging some of the core elements at the expense of others and replacing the latter by the peripheral components. Second, we suggest that the four ways of handling the compatibility gap between a managerial technique and the context of its implementation, namely transformation, customisation, loose-coupling and corruption (Lozeau et al. 2002), can represent the stages of the same process rather than distinct independent categories. Finally, we demonstrate that the eroding effect of macro-level institutional arrangements on sustainable service improvement is realised through the duality and inequality of its goals, whereby (implicit) long-term sustainability-related objectives become marginalised in favour of (explicit) short-term target-driven performance objectives.

Our analysis is likely to be applicable to a wide range of theoretically informed managerial techniques and service improvement approaches deployed by public sector organisations, particularly those relying on purposefully created roles to enable organisational change and embed it in social structures. It also emphasises the lack of attention to the sustainability of improvement in the current policy context, which can result in a failure to sustain the outcomes of improvement once they have been attained, measured and reported. Finally, we highlight the need to be conscious of the fine line between the *adaptation* of managerial techniques advocated in the managerialist literature and their *corruption* with an associated loss of potential to enact and sustain positive change. An exploration of new ways of maintaining context-sensitive customisation of improvement techniques without slipping into loose-coupling and corruption could provide a useful direction for future empirical enquiry.

References

Addicott, R., McGivern, G., & Ferlie, E. (2006). Networks, organizational learning and knowledge management: NHS cancer networks. *Public Money & Management, 26*(2), 87–94.

Addicott, R., McGivern, G., & Ferlie, E. (2007). The distortion of a managerial technique? The case of clinical networks in UK health care. *British Journal of Management, 18*(1), 93–105.

Baskerville, N. B., Liddy, C., & Hogg, W. (2012). Systematic review and meta-analysis of practice facilitation within primary care settings. *Annals of Family Medicine, 10*(1), 63–74.

Bate, S. P., & Robert, G. (2002). Knowledge management and communities of practice in the private sector: Lessons for modernizing the National Health Service in England and Wales. *Public Administration, 80*(4), 643–663.

Berta, W., Cranley, L., Dearing, J. W., Dogherty, E. J., Squires, J. E., & Estabrooks, C. A. (2015). Why (we think) facilitation works: Insights from organizational learning theory. *Implementation Science, 10*, 141.

Boaden, R., Harvey, G., Moxham, C., & Proudlove, N. (2008). *Quality improvement: Theory and practice in healthcare.* Coventry: NHS Institute for Innovation and Improvement.

Bosch, M., Van Der Weijden, T., Wensing, M., & Grol, R. (2007). Tailoring quality improvement interventions to identified barriers: A multiple case analysis. *Journal of Evaluation in Clinical Practice, 13*(2), 161–168.

Causer, G., & Exworthy, M. (1998). Professionals as managers across the public sector. In M. Exworthy & S. Halford (Eds.), *Professionals and the new managerialism in the public sector* (pp. 83–101). Buckingham: Open University Press.

Currie, G. (2007). Spanning boundaries in pursuit of effective knowledge sharing within networks in the NHS. *Journal of Health Organization and Management, 21*(4/5), 406–417.

Currie, G., & Suhomlinova, O. (2006). The impact of institutional forces upon knowledge sharing in the UK NHS: The triumph of professional power and the inconsistency of policy. *Public Administration, 84*(1), 1–30.

Damschroder, L. J., Aron, D. C., Keith, R. E., Kirsh, S. R., Alexander, J. A., & Lowery, J. C. (2009). Fostering implementation of health services research findings into practice: A consolidated framework for advancing implementation science. *Implementation Science, 4,* 50.

Denis, J. L., Hebert, Y., Langley, A., Lozeau, D., & Trottier, L. H. (2002). Explaining diffusion patterns for complex health care innovations. *Health Care Management Review, 27*(3), 60–73.

Dixon-Woods, M., McNicol, S., & Martin, G. (2012). Ten challenges in improving quality in healthcare: Lessons from the Health Foundation's programme evaluations and relevant literature. *BMJ Quality & Safety, 21*(10), 876–884.

Dogherty, E. J., Harrison, M. B., & Graham, I. D. (2010). Facilitation as a role and process in achieving evidence-based practice in nursing: A focused review of concept and meaning. *Worldviews on Evidence-Based Nursing, 7*(2), 76–89.

Fervers, B., Burgers, J. S., Haugh, M. C., Latreille, J., Mlika-Cabanne, N., Paquet, L., Coulombe, M., Poirier, M., & Burnand, B. (2006). Adaptation of clinical guidelines: Literature review and proposition for a framework and procedure. *International Journal for Quality in Health Care, 18*(3), 167–176.

Harrison, S., & Pollitt, C. (1994). *Controlling health professionals: The future of work and organization in the National Health Service.* Buckingham: Open University Press.

Harvey, G., & Kitson, A. (Eds.). (2015). *Implementing evidence-based practice in healthcare: A facilitation guide.* London: Routledge.

Harvey, G., Kitson, A., & Munn, Z. (2012). Promoting continence in nursing homes in four European countries: The use of PACES as a mechanism for improving the uptake of evidence-based recommendations. *International Journal of Evidence-Based Healthcare, 10*(4), 388–396.

Harvey, G., Loftus-Hills, A., Rycroft-Malone, J., Titchen, A., Kitson, A., McCormack, B., & Seers, K. (2002). Getting evidence into practice: The role and function of facilitation. *Journal of Advanced Nursing, 37*(6), 577–588.

Kamoche, K., Kannan, S., & Siebers, L. Q. (2014). Knowledge-sharing, control, compliance and symbolic violence. *Organization Studies, 35*(7), 989–1012.

Kelly, D., Simpson, S., & Brown, P. (2002). An action research project to evaluate the clinical practice facilitator role for junior nurses in an acute hospital setting. *Journal of Clinical Nursing, 11,* 90–98.

Kirsh, S. R., Lawrence, R. H., & Aron, D. C. (2008). Tailoring an intervention to the context and system redesign related to the intervention: A case study of implementing shared medical appointments for diabetes. *Implementation Science, 3*, 34.

Kislov, R., Harvey, G., & Walshe, K. (2011). Collaborations for leadership in Applied Health Research and Care: Lessons from the theory of communities of practice. *Implementation Science, 6*, 64.

Kislov, R., Hodgson, D., & Boaden, R. (2016). Professionals as knowledge brokers: The limits of authority in healthcare collaboration. *Public Administration, 94*(2), 472–489.

Kislov, R., Walshe, K., & Harvey, G. (2012). Managing boundaries in primary care service improvement: A developmental approach to communities of practice. *Implementation Science, 7*, 97.

Kislov, R., Waterman, H., Harvey, G., & Boaden, R. (2014). Rethinking capacity building for knowledge mobilisation: Developing multilevel capabilities in healthcare organisations. *Implementation Science, 9*, 166.

Kitchener, M. (2000). The 'bureaucratization' of professional roles: The case of clinical directors in UK hospitals. *Organization, 7*(1), 129–154.

Kitson, A., Harvey, G., & McCormack, B. (1998). Enabling the implementation of evidence based practice: A conceptual framework. *Quality & Safety in Health Care, 7*(3), 149–158.

Kitson, A. L., Rycroft-Malone, J., Harvey, G., McCormack, B., Seers, K., & Titchen, A. (2008). Evaluating the successful implementation of evidence into practice using the PARiHS framework: Theoretical and practical challenges. *Implementation Science, 3*, 1.

Krein, S. L., Damschroder, L. J., Kowalski, C. P., Forman, J., Hofer, T. P., & Saint, S. (2010). The influence of organizational context on quality improvement and patient safety efforts in infection prevention: A multicenter qualitative study. *Social Science and Medicine, 71*(9), 1692–1701.

Langley, A. (1999). Strategies for theorizing from process data. *Academy of Management Review, 24*(4), 691–710.

Lewis, L. K., & Seibold, D. R. (1993). Innovation modification during intraorganizational adoption. *Academy of Management Review, 18*(2), 322–354.

Lozeau, D., Langley, A., & Denis, J.-L. (2002). The corruption of managerial techniques by organizations. *Human Relations, 55*(5), 537–564.

McCann, L., Hassard, J. S., Granter, E., & Hyde, P. J. (2015). Casting the lean spell: The promotion, dilution and erosion of lean management in the NHS. *Human Relations, 68*(10), 1557–1577.

Mintzberg, H., & Waters, J. A. (1985). Of strategies, deliberate and emergent. *Strategic Management Journal, 6*(3), 257–272.

Nadin, S., & Cassell, C. (2004). Using data matrices. In C. Cassell & G. Symon (Eds.), *Essential guide to qualitative methods in organizational research* (pp. 271–287). London: SAGE Publications.

Omidvar, O., & Kislov, R. (2016). R&D consortia as boundary organisations: Misalignment and asymmetry of boundary management. *International Journal of Innovation Management, 20*(2), 1650030.

Petrova, M., Dale, J., Munday, D., Koistinen, J., Agarwal, S., & Lall, R. (2010). The role and impact of facilitators in primary care: Findings from the implementation of the Gold Standards Framework for Palliative Care. *Family Practice, 27*, 38–47.

Powell, A. E., & Davies, H. T. O. (2012). The struggle to improve patient care in the face of professional boundaries. *Social Science and Medicine, 75*(5), 807–814.

Radnor, Z. J., Holweg, M., & Waring, J. (2012). Lean in healthcare: The unfilled promise? *Social Science and Medicine, 74*(3), 364–371.

Ruhe, M. C., Carter, C., Litaker, D., & Stange, K. C. (2009). A systematic approach to practice assessment and quality improvement intervention tailoring. *Quality Management in Healthcare, 18*(4), 268–277.

Seers, K., Cox, K., Crichton, N. J., Edwards, R. T., Eldh, A. C., Estabrooks, C. A., Harvey, G., Hawkes, C., Kitson, A., & Linck, P. (2012). FIRE (Facilitating Implementation of Research Evidence): A study protocol. *Implementation Science, 7*, 25.

Shipman, C., Addington-Hall, J., Thompson, M., Pearce, A., Barclay, S., Cox, I., Maher, J., & Millar, D. (2003). Building bridges in palliative care: Evaluating a GP Facilitator programme. *Palliative Medicine, 17*(7), 621–627.

Stetler, C., Legro, M., Rycroft-Malone, J., Bowman, C., Curran, G., Guihan, M., Hagedorn, H., Pineros, S., & Wallace, C. (2006). Role of "external facilitation" in implementation of research findings: A qualitative evaluation of facilitation experiences in the Veterans Health Administration. *Implementation Science, 1*, 23.

Tierney, S., Kislov, R., & Deaton, C. (2014). A qualitative study of a primary-care based intervention to improve the management of patients with heart failure: The dynamic relationship between facilitation and context. *BMC Family Practice, 15*, 153.

16

Stakeholders' Involvement and Service Users' Acceptance in the Implementation of a New Practice Guideline

Comfort Adeosun, Lorna McKee and Hilary Homans

Introduction

Background

The adoption and implementation of clinical guidelines has a positive impact on quality, service effectiveness and patient care (David and Taylor-Vaisey 1997; Grimshaw et al. 2004). However, implementing evidence-based practice and practice guidelines is complex and challenging (Taylor et al. 2011). These challenges have been shown to range from individual provider behaviour, quality and characteristics of the guidelines, patient characteristics to organisational characteristics, settings and health system-level factors (David and Taylor-Vaisey 1997; Greenhalgh et al. 2004; Francke et al. 2008; Urquhart et al. 2014). One way of surfacing the factors that impact on the uptake and effective implementation of clinical guidelines is to undertake comparative

C. Adeosun (✉) · L. McKee · H. Homans
University of Aberdeen, Aberdeen, UK
e-mail: Comfort.adeosun.08@aberdeen.ac.uk

© The Author(s) 2018
A.M. McDermott et al. (eds.), *Managing Improvement in Healthcare*,
Organizational Behaviour in Health Care,
https://doi.org/10.1007/978-3-319-62235-4_16

studies of the same guideline as it is rolled out in different contexts (Yin 2003; Helfrich et al. 2007).

There is currently strong evidence that both the internal and external context of the organisation influence the implementation and utilisation of guidelines, confirming that implementation processes are complex, interactive and iterative in nature (Johns 2001, 2006; Fitzgerald et al. 2002; Krein et al. 2010; McDermott and Keating 2012). It has been further suggested that to ensure successful implementation, appropriate customised implementation policies and practices must be deployed by local healthcare organisations (Klein and Sorra 1996; Weiner et al. 2008). Increasingly, it is argued that there is a need to develop an in-depth appreciation of the formation and role of these localised implementation policies and practices.

According to Helfrich et al. (2007), implementation policies and practices are 'the formal strategies (that is, the policies) the organisation uses to put the innovation into use and the actions that follow from those strategies (that is, practices)' (p. 284). Some of these strategies may include: the quality and quantity of training; rewards, including promotion, incentives, praise or improved working conditions; effective communication about the goals of the implementation; sufficient time for users to experiment or learn new skills related to the innovation; and the quality, accessibility and user-friendliness of the innovation itself (Helfrich et al. 2007; Weiner et al. 2008). An organisation's implementation policies and practices will influence innovation implementation and use by shaping the organisation's implementation climate (Klein and Sorra 1996) irrespective of the type of guideline. However, the importance of stakeholder and especially patient involvement in implementation and improvement activities should not be overlooked.

Whilst the need for patient involvement or participation in quality improvement activities is increasingly gaining attention (Donetto et al. 2014b; Vahdat et al. 2014; Wiig et al. 2014), within the organisational literature, there is only a limited number of studies which capture the impact and significance of stakeholders and in particular patient involvement in implementation and improvement (Damschroder et al. 2009; Urquhart et al. 2014). In this chapter, we address this acute gap in knowledge and use the broad concept of stakeholders to include the whole mix of healthcare providers, policymakers, as well as end users, patient groups, the public and funders. The new insights added by this chapter include the influence of stakeholders, end users and the community on the complex implementation of a new guideline dependent on context and system-level factors.

This focus on patients and end users is timely because patient involvement in healthcare decision-making is becoming increasingly promoted as an important tool for improving quality of care (Parsons et al. 2010; Vahdat et al. 2014). It is suggested that the more patients are involved, the more they can help to co-design their care and improve it. Recent studies have been conducted using the experience-based co-design method in healthcare improvement (Donetto et al. 2014a, b; Locock et al. 2014). The method encourages staff, patients and carers to reflect on their experience of care and look for ways to improve the process and assess the achievements of any changes implemented (Donetto et al. 2014b). Other methods include patient participation and shared decision-making (Wiig et al. 2014), and the shared ambition is to create services and innovations that are as 'user- and carer-led as possible' (Sheldon and Harding 2010, p. 5).

Focused Antenatal Care (FANC) Model

In this chapter, the focus is on the implementation of the Focused Antenatal Care (FANC) model, a clinical practice guideline developed by the World Health Organisation (WHO) to improve the quality of antenatal care (WHO 2006). In 2007, the Federal Ministry of Health in Nigeria adopted the WHO standards and guidelines to improve maternal, neonatal and child health. The available evidence suggests incomplete and weak implementation of the FANC model in Nigeria (FMOH 2011). In general, the coverage and content of care provided during antenatal care are regarded as sub-optimal across the nation (Osungbade et al. 2008; Okoli et al. 2012). This chapter addresses this puzzle as to why such variation occurs, both drawing on the need to understand local implementation policies and practices as well as focusing on how stakeholders and end users impact implementation and uptake. In this chapter, terms such as the successful implementation or effectiveness of the intervention, the FANC model, are used interchangeably with intervention or innovation use. They are all used to mean the committed, consistent and routine utilisation of the new practice guideline in the organisations studied.

The overall study upon which this chapter is based draws upon a theoretical framework first developed by Klein and Sorra (Klein and Sorra 1996; Klein et al. 2001) and later revised by Helfrich et al. (2007). The theory suggests that the presence or absence of factors such as management support, resources and appropriate implementation policies and practices can facilitate or hinder the successful implementation of innovation. As an organisational-level framework, it is concerned with innovations requiring coordinated use by multiple organisational members. The model was adapted to accommodate other factors from the extant literature. Significantly, the context of the healthcare organisation and system-level factors are stressed as important influencing factors (Johns 2006; McDermott and Keating 2012; Urquhart et al. 2014). The adapted framework is shown in Fig. 16.1.

Methods

The overall study adopted a case study research methodology (Creswell 2007; Yin 2003). It employed a multi-method qualitative approach to data collection which is both descriptive and exploratory in nature (Patton 1990; Fitzgerald and Dopson 2009). Four comparative case studies in one state in the Niger Delta area of Nigeria were purposively chosen and provided an opportunity for contemporaneous study of implementation of FANC in a range of diverse local contexts.

Four healthcare settings were selected across the three levels of healthcare provision in the state (tertiary, secondary and primary healthcare levels) based on the levels of ownership (private and public) and teaching status (teaching and non-teaching). Attempts were also made to access secondary data from the local and state governments on the

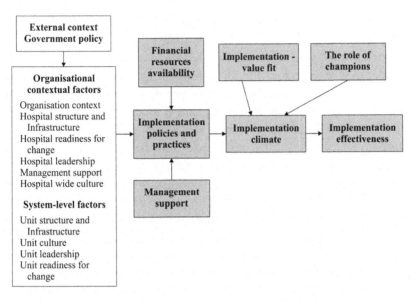

Fig. 16.1 Theoretical model for complex innovation implementation. Adapted from Klein and Sorra 1996; Helfrich et al. 2007. *Shaded* original model. *Unshaded* proposed extension to model

adoption and utilisation of the FANC guideline by the selected hospitals and healthcare facilities.

The overall study protocol summary, consent form and ethics approval form were approved by the University of Aberdeen College of Arts and Social Sciences Research Ethics and Governance Committee. Subsequent ethical approvals were obtained from the Niger Delta State Primary Health Care Management Board, and from each case study site management team.

Three predesigned research instruments were used to collect data from the healthcare providers and policymakers, and importantly it was an explicit objective of the study to gather data directly from the end users of the services. A sample of pregnant women across the four case study sites was interviewed about their perceptions and knowledge of the FANC initiative and its goals. For the providers, the interview and focus group discussion schedule included questions on the adoption/adaptation and implementation of the FANC model. The schedule also included questions on how the model guideline was generally perceived and promoted in the organisation, amongst other issues surrounding improving the quality of antenatal care. The questions were open-ended in order to give room for other themes to emerge during the interview.

For the pregnant women, the interview and focus group discussion guide included topics such as the pregnant woman's gestational age, number of antenatal care visits, awareness and perception of the FANC model, information received during health talks and the intention to deliver in the facility. The interviews lasted between forty and ninety minutes.

The prescribed FANC guideline checklists were also used to obtain data during substantial time spent in non-participant observation of antenatal care clinic sessions. The data were collected from January to May 2013. In order to assess the factors and variables of interest, different cadres of staff were invited to participate, and the pregnant women also came from different backgrounds in terms of age, number of previous pregnancies, proximity to the facility and health status. Appropriate secondary data and policy documents related to the implementation of the new guideline were collected from each facility's medical records.

A thematic framework analysis (Ritchie et al. 2014) was used to analyse the qualitative data. Data analysis began when the first data were collected. Codes and categories were generated from the data using inductive and deductive approaches (Guest et al. 2012). The framework was flexible, and codes and themes were reassessed as new codes or themes emerged. The coding process was guided by the coding principles proposed by Braun and Clarke (2006) and Ritchie et al. (2014). Each code contains evidence from the manuscripts with links to the data. The *NVivo 10* qualitative data management software was used to support the analysis stage. The secondary data were obtained from document reviews on antenatal clinic utilisation, and checklists were analysed using descriptive statistics. Data from each case study site were analysed separately before comparison with other study sites. The results were integrated and triangulated at the data analysis and interpretation stage (Bryman 2006). A stage-by-stage data analysis and triangulation helped to gain deeper meanings and insights into the implementation

Table 16.1 Total number of participants, role and facility

Data sources		Case A Comprehensive health centre	Case B Tertiary hospital	Case C Primary health-care	Case D Private hospital
Interviews					
Providers	Antenatal care staff	3	4	3	3
	Management level staff	5	3	2	4
Policymakers		1	2	2	2
Pregnant women (Includes first visit and revisits)		9	10	8	6
Total interviews		18	19	15	13
Focus group discussion					
	Providers	1(n = 5)	1(n = 5)	1(n = 5)	
	Pregnant women	3(n = 16)	2(n = 11)	1(n = 5)	2(n = 10)
Non-participant observation (hours)		124	80	45	40

Source Authors

process in each local context. The cross-case analysis was conducted to enhance the findings' generalisability or transferability to other contexts and deepen understanding and explanation of the phenomenon being studied (Fitzgerald and Dopson 2009; Miles et al. 2014; Yin 2003). The analysis compared and contrasted themes between and within the case study sites. The following table summarises the participants in the study (Table 16.1).

Case Description

Case A is a public and comprehensive healthcare facility. It provides general outpatient care, maternal and child healthcare services, amongst others. The facility is fully implementing the FANC guideline as recommended by the WHO. It has a community health insurance scheme for service users.

Case B is a tertiary and teaching hospital. The hospital has a local protocol for antenatal care similar in content to the FANC guideline. The facility is not implementing the FANC model. The antenatal care visits follow the traditional model with ten to twelve visits in one pregnancy.

Case C is a public primary healthcare centre funded by the state ministry of health. It receives supportive supervision from the primary healthcare board. The facility is partially implementing the FANC model due to pregnant women's rejection of a reduction in the number of antenatal care visits. Free medical care is provided by the state government.

Case D is a private and non-teaching hospital providing primary and secondary care. Most of the pregnant women receiving care in this facility are graduates. The content of the antenatal care is incongruent with the FANC guideline. For financial reasons, the number of antenatal care visits follows the traditional model. Notably, pregnant women pay for consultation each visit.

Findings

In the following section, we present the findings on the FANC guideline implementation. In particular, we aim to show the complex interplay of different levels of influence in each case from local policy, and providers' adaptive behaviours to the local pregnant women's action on implementation policies and practices. Notably, the implementation team members were prescribed in the implementing facilities. They were responsible for fulfilling specific roles in line with the new guideline in each organisation. The impact of this on the implementation process is that each facility had to demonstrate that they were aiming to comply with the government policy on the FANC model as a top-down strategy. However, the private hospital implemented the FANC model in response to the need for evidence-based practice.

Table 16.2   Cross-case matrix: implementation policies and practices

Implementation policies and practices	Case A	Case B	Case C	Case D
Training (on prevailing antenatal care practice)	Y	Y	Y	Y
Communication of FANC model	Y	-	Y	Y
72-hour roster	-	-	Y	-
Community involvement and engagement	Y	-	Y	-
Employment of key staff	Y	-	Y	Y
Adaptation and innovation to FANC	Y	-	Y	Y
Innovation in antenatal care, e.g. health insurance scheme, multiple informants for health talk	Y	Y	Y	Y
Protocol or Standard Operating Procedures	Y	Y	-	Y
Audit and feedback mechanisms	Y	Y	Y	Y

Y = Present; - = Absent
Source Authors

Implementation Policies and Practices (IPPs)

The cross-case analysis revealed various effective implementation policies and practices (IPPs) that affected the FANC model (or the local protocol) implementation and routine utilisation across the cases. The IPPs adopted across cases to facilitate implementation are shown in Table 16.2. The findings are divided into common and distinctive IPPs.

Common and Effective Implementation Policies and Practices: Similarities Across Cases

The data revealed *three* similar implementation practices across all cases, as described below.

Training

The four facilities provided training for their staff. Participants from Cases A and C reported that the local government organised training for staff on the FANC model guideline at the start of the implementation process. The adoption of the FANC model as a government policy meant that healthcare facilities were mandated to accept and use it. The training created the awareness and knowledge needed by staff to provide care in line with the new model.

> The Focused Antenatal Care model was introduced to us as a policy. A workshop was organised and the concept of Focused Antenatal Care was explained to us. From there we started the implementation. [State Reproductive Health Manager, Board]

Training was also organised for antenatal care staff at Case D, the private hospital, in order to embrace the WHO best practice for quality antenatal care.

Innovation in Antenatal Care Practices

Numerous innovative ways were used to support the implementation process. In Case A, the introduction of the Community Health Insurance Scheme enhanced the implementation process. The scheme is perceived to be one of the key facilitators for implementation and continuous utilisation of the FANC guideline in the hospital.

> The Community Health Insurance Scheme, I will say, is one of the major facilitating factors. Because I know that Focused Antenatal Care, even in other health facilities like the primary health centres, ought to use Focused Antenatal Care, but most of them you can see that mothers are not embracing Focused Antenatal Care there. [Case A, Senior Manager 2]

Each facility engaged multiple informants for health talk on antenatal care clinic days. At Case B, different healthcare professionals presented health talks. This included antenatal care staff, family planning consultants and physiotherapists.

Audit and Feedback Mechanisms

Audit and feedback tools were used as a mechanism to monitor and evaluate FANC model use in Cases A and C. In Case B, regular clinical audits were conducted in addition to Tuesday weekly clinic meetings. Also, in Cases B and D, feedback tools were used to monitor performance and indicated needed improvements in the quality of care being provided to pregnant women.

Distinctive Effective Implementation Policies and Practices Across Cases

Distinctive policies and workplace practices were also observed in the four facilities. Table 16.2 above shows that Cases A and C engaged several implementation policies and practices to implement the FANC model. Also, both facilities were under the state government supervision mechanism (a form of top-down strategy for policy implementation).

Community Engagement and Involvement

Community engagement and involvement were embarked upon by the management in Cases A and C. Involving the local community chiefs to communicate about the importance of antenatal care to reduce maternal mortality, and where the facilities were situated, may have facilitated community ownership and patronage. This is in line with the model's recommendation (WHO 2006).

> The impact on the antenatal programme has been positive. When it came we went to the paramount Ruler who is the custodian of this place and told him of the new model. He mandated his town crier to take the announcement round the community. And that was the first step in the initial enlightenment campaign. [Case C, Senior Manager, Doctor]

The community chiefs and religious leaders play an important role within the community. Many times matters of faith and tradition conflict with conventional medicine; hence, the need to engage the leaders to understand the importance of quality antenatal care. This strategy proved to be effective as many women visited the facilities for antenatal care.

Staff Employment

The data revealed the employment of key staff to support the model implementation in three facilities—Cases A, C and D.

Communication of FANC Model by Appropriate Authorities

The communication of the FANC model and training received by providers at Cases A, C and D are perceived to be key implementation practices in gaining the support of staff to use the guideline. This further shows that the structure of the healthcare system and the management processes in each facility affected the effectiveness of guideline implementation.

> We get information from the Western world and we want to see how we can improve. Through adaptation, we want our people to get the best so we have to improve on our own and on the knowledge we have. It is done worldwide; why should we be left behind? It is the drive to get evidence-based practice into the system. [Case D, Senior Consultant, Obstetrician and Gynaecologist]

Staff Adaptive Innovative Behaviour

The adoption of a seventy-two-hour shift by the midwives at Case C was exceptional. It was one of the internal implementation policies put in place to ensure that the model was implemented as a state policy.

Without this strategic action, pregnant women may not have reported for care if they were unsure that staff would be available to care for them.

The Use of Standard Operating Procedures and Protocols

The use of appropriate antenatal care protocols and appropriate staff at Cases B and D appeared to have had an impact on the implementation of the antenatal care protocol in use in each facility. The hospitals' compliance to protocol utilisation was assessed through observation and interview.

Adaptation and Innovation to FANC

Many pregnant women in Nigeria seek care exclusively in the church or with traditional birth attendants because they believe that through prayers and sometimes with traditional medicine, complications leading to Caesarean sections may be averted. Responding to this challenge, many providers now invite religious leaders to incorporate prayer sessions into the antenatal care schedule to encourage attendance and willingness to deliver with the aid of skilled birth attendants.

A pregnant woman summarised the antenatal care clinic at Case C thus:

> The first thing we do is to pray. After that they preach. This is followed by the health talk and after the health talk we get our folders, then we go upstairs for our laboratory test. [Case C, PW 6]

Due to the rejection of the FANC model's recommended four antenatal care visits, the providers at Cases A and C encouraged pregnant women to visit the facility when they were sick. This was aimed at discouraging pregnant women from using mission homes, faith-based organisations and traditional birth attendants. As a result, more women embraced the new guideline in these facilities.

> They [midwives] said we should come four times during our pregnancy
> ... If they check you and everything about you is okay, your visit here is
> to be four times, depending on the condition of the pregnancy. They said
> if you have any complication you can come before the date given to you.
> [Case A, PW 7]

In addition, pregnant women receiving care at Case C were offered free medical care in order to encourage antenatal care attendance. This suggests that free medical care is a key contextual factor facilitating the implementation of the FANC model in the state.

Despite pregnant women's refusal of the reduction in the number of antenatal care visits, these implementation strategies employed by the facilities to encourage guideline use in the facility and win pregnant women's trust helped to facilitate continuous innovation use. These adaptive behaviours influenced the implementation climate in each case study site. When these distinctive factors are linked together with reports from pregnant women (these findings are reported in another chapter), it appears that the ongoing effective implementation of the FANC model at the public facilities, particularly Cases A and C, are due to the support received from the state and local governments and community involvement. The providers' response to service users' preference for antenatal care boosted implementation efforts.

It appears that there are diverse interpretations of what constitutes successful or effective implementation of the FANC model for the various actors in each facility. For the pregnant women, it was their ability to visit the antenatal care clinic frequently in defiance of the optimal number of visits laid out in the policy guideline. For the providers, effective implementation meant providing quality antenatal care despite limited resources in line with the new guideline. At the private hospital, more visits meant more money and profit maximisation. The policymakers and the local chiefs perceived effectiveness as the increase in the total number of pregnant women accessing antenatal care in the healthcare facilities with the aim of reducing overall maternal and infant mortality in their communities and the state at large. The religious leaders perceived effectiveness as supporting more pregnant women in receiving conventional care in addition to prayers and faith.

Discussion

This chapter has examined the implementation policies and practices that influenced the implementation of the FANC model in four healthcare settings. The findings revealed three common IPPs across the cases studied—training, innovation in antenatal care practices, and audit and feedback mechanisms. Distinctive IPPs were observed in the four cases. Interestingly, community involvement and engagement prompted other practices observed in two cases. Stakeholders' involvement and service users' acceptance/resistance led to staff adaptive innovative behaviour and adaptation to the FANC model implementation. These findings showed that external and internal organisational context and the healthcare system influenced the implementation policies and practices engaged. The data indicate that a range of different contextual factors and internal policies interacted to facilitate implementation, as also observed in other studies (Dixon-Woods 2014; Fitzgerald et al. 2002; Hovlid and Bukve 2014).

As stated earlier, two of the four cases, Cases A and C, demonstrated the importance of community engagement and stakeholder involvement in innovation/intervention implementation. The findings showed that the increase in the number of attendees and improvement in the facilities were the result of collaboration between the healthcare organisation, the community and religious leaders and other stakeholders. The importance of stakeholder involvement has been documented in other healthcare studies (Damschroder et al. 2009; Hovlid and Bukve 2014; Urquhart et al. 2014). However, the interaction between them in this study on FANC model implementation was a unique finding.

In addition to stakeholders' involvement, the findings show that service users' (pregnant women's) acceptance or resistance to the FANC model had an impact on implementation effectiveness. Pregnant women's perception of care and dislike of the reduced number of visits as recommended in the model influenced the organisational responses and providers' implementation efforts. This is a contextual influence neglected in previous studies (Johns 2006; Pettigrew et al. 1992). This factor influenced implementation practices in Cases A and B. For example, the providers' inclusion of prayer into the antenatal care practice was in response to service users' religious beliefs and the importance attached to prayer. Also, the sociocultural influence of the traditional birth attendant on pregnant women's health-seeking behaviour generated varied responses from each organisation. All these demonstrate that external contextual factors, including service users' acceptance, influence implementation of evidence-based practice in healthcare facilities. This creates the need for patient involvement in improving implementation efforts and should go beyond involvement in guideline development (Sheldon and Harding 2010; Wiig et al. 2014).

It also indicates that end users or service users are not passive in the implementation process. They are active change agents to co-shape implementation effectiveness together with management support, model champions, community and stakeholders. A new publication by the WHO has increased the recommended number of visits to eight (WHO 2016).

Implications

The chapter shows the need for stakeholder and patient involvement in the adoption of new innovations or interventions. The purpose and the expected outcomes of interventions should be explained in order for end users—including practitioners and patient groups—to express their opinions about changes that may be explored as a result of the new interventions. Practitioners/providers should continue to provide evidence-based practice.

Conclusions

This chapter contributes to and fills the research-knowledge gap and evidence-based practice-implementation gap in the implementation of a maternal health clinical practice guideline in Nigeria. A major finding is the broader nature and extent of external context in guideline implementation. It showed that stakeholder involvement, the role of wider community involvement and service users' acceptance/resistance all influenced implementation climate and effectiveness. It also affected the continuous innovation or intervention use.

References

Braun, V., & Clarke, V. (2006). Using thematic analysis in psychology. *Qualitative Research in Psychology, 3*(2), 77–101.

Bryman, A. (2006). Integrating quantitative and qualitative research: How is it done? *Qualitative Research, 6*(1), 97–113.

Creswell, J. W. (2007). *Qualitative inquiry and research design: The five approaches.* Thousand Oaks: Sage.

Damschroder, L. J., Banaszak-Holl, J., Kowalski, C. P., Forman, J., Saint, S., & Krein, S. L. (2009). The role of the 'champion' in infection prevention: Results from a multisite qualitative study. *Quality and Safety in Health Care, 18*(6), 434–440.

David, A. D., & Taylor-Vaisey, A. (1997). Translating guidelines into practice: A systematic review of theoretic concepts, practical experience and research evidence in the adoption of clinical practice guidelines. *Journal of Canadian Medical Association, 157*(4), 408–416.

Dixon-Woods, M. (2014). *Perspectives on context: The problems of context in quality improvement.* London: The Health Foundation.

Donetto, S., Tsianakas, V., & Robert, G. (2014a). *Using Experience-based Co-design to improve the quality of healthcare: Mapping where we are now and establishing future directions.* London: King's College London.

Donetto, S., Pierri, P., Tsianakas, V., & Robert, G. (2014b). *Experience-based Co-design and health care improvement: Realising participatory design in the public sector.* Fourth Service Design and Innovation Conference. Retrieved from http://www.servdes.org/wp/wp-content/uploads/2014/06/Donetto-S-Pierri-P-Tsianakas-V-Robert-G.pdf. Accessed 22 Feb 2016.

Federal Ministry of Health. (2011). *Saving newborn lives in Nigeria: Newborn health in the context of the Integrated Maternal, Newborn and Child Health Strategy* (2nd ed.). Abuja: Federal Ministry of Health, Save the Children, Jhpiego.

Fitzgerald, L., & Dopson, S. (2009). Comparative case study designs: Their utility and development. In A. Bryman & D. A. Buchannan (Eds.),

Organisational research in the SAGE handbook of organisational research methods. London: Sage.

Fitzgerald, L., Ferlie, E., Wood, M., & Hawkins, C. (2002). Interlocking interactions, the diffusion of innovations in health care. *Human Relations, 55*(12), 1429–1449.

Francke, A., Smit, M., de Veer, A., & Mistiaen, P. (2008). Factors influencing the implementation of clinical guidelines for health care professionals: A systematic meta-review. *BMC Medical Informatics and Decision Making, 8*(1), 38.

Greenhalgh, T., Robert, G., Macfarlane, F., Bate, P., & Kyriakidou, O. (2004). Diffusion of innovations in service organisations: Systematic review and recommendations. *Milbank Quarterly, 82*(4), 581–629.

Grimshaw, J., Eccles, M., & Tetroe, J. (2004). Implementing clinical guidelines: Current evidence and future implications. *Journal of Continuing Education for Health Professions, 24*(Suppl. 1), S31–S37.

Guest, G., MacQueen, K. M., & Namey, E. M. (2012). *Applied thematic analysis*. London: Sage.

Helfrich, C. D., Weiner, B. J., McKinney, M. M., & Minasian, L. (2007). Determinants of implementation effectiveness: Adapting a framework for complex innovations. *Medical Care Research and Review, 64*(3), 279–303.

Hovlid, E., & Bukve, O. (2014). A qualitative study of contextual factors' impact on measures to reduce surgery cancellations. *BMC Health Services Research, 14*(1), 215.

Johns, G. (2001). In praise of context. *Journal of Organisational Behavior, 22*(1), 31–42.

Johns, G. (2006). The essential impact of context on organizational behavior. *Academy of Management Review, 31,* 396–408.

Klein, K. J., & Sorra, J. S. (1996). The challenge of innovation implementation. *The Academy of Management Review, 21*(4), 1055–1080.

Klein, K. J., Conn, A. B., & Sorra, J. S. (2001). Implementing computerised technology: An organisational analysis. *Journal of Applied Psychology, 86*(5), 811–824.

Krein, S. L., Damschroder, L. J., Kowalski, C. P., Forman, J., Hofer, T. P., & Saint, S. (2010). The influence of organisational context on quality improvement and patient safety efforts in infection prevention: A multicenter qualitative study. *Social Science and Medicine, 71*(9), 1692–1701.

Locock, L., Robert, G., Boaz, A., Vougioukalou, S., Shuldam, C., Fielden, J., Ziebland, S., Gager, M., Tollyfiled, R., & Pearcey, J. (2014). Testing accelerated experience-based co-design: A qualitative study of using a national archive of patient experience narrative interviews to promote rapid patient-centred service improvement. *Health Services and Delivery Research, 2*(4). doi:10.3310/hsdr02040.

McDermott, A. M., & Keating, M. A. (2012). Making service improvement happen: The importance of social context. *The Journal of Applied Behavioural Science, 48*(1), 62–92.

Miles, M. B., Huberman, A. M., & Saldaña, J. (2014). *Qualitative data analysis and methods sourcebook* (3rd ed.). London: Sage.

Okoli, U., Abdullahi, M. J., Pate, M. A., et al. (2012). Prenatal care and basic emergency obstetric care services provided at primary healthcare facilities in rural Nigeria. *International Journal of Gynecology & Obstetrics, 117*(1), 61–65.

Osungbade, K., Oginni, S., & Olumide, A. (2008). Content of antenatal care services in secondary health care facilities in Nigeria: Implication for quality of maternal health care. *International Journal for Quality in Health Care, 20*(5), 346–351.

Parsons, S., Winterbottom, A., Gross, P., & Redding, D. (2010). *The quality of patient engagement and involvement in primary care.* London: The King's Fund.

Patton, M. Q. (1990). *Qualitative evaluation and research method* (2nd ed.). London: Sage.

Pettigrew, A., Ferlie, E., & McKee, L. (1992). Shaping strategic change—The case of the NHS in the 1980s. *Public Money & Management, 12*(3), 27–31.

Ritchie, J., Lewis, J., Nicholis, C. M., & Ormston, R. (Eds.). (2014). *Qualitative research practice: A guide for social science students & researchers.* London: Sage.

Sheldon, K., & Harding, E. (2010). *Good practice guidelines to support the involvement of service users and carers in clinical psychology services.* Retrieved from http://www.ispraisrael.org.il/Items/00289/Service_user_and_carer_involvement.pdf. Accessed 22 Feb 2016.

Taylor, S. L., Dy, S., Foy, R., Hempel, S., McDonald, K. M., Øvretveit, J., Pronovost, P. J., Rubenstein, L. V., Wachter, R. M. & Shekelle, P. G. (2011). What context features might be important determinants of the effectiveness of patient safety practice interventions? *BMJ Quality & Safety, 20*(7), 611–617.

Urquhart, R., Porter, G. A., Sargeant, J., Jackson, L., & Grundfeld, E. (2014). Multi-level factors influence the implementation and use of complex innovations in cancer care: A multiple case study of synoptic reporting. *Implementation Science, 9*(1), 121.

Vahdat, S., Hamzehgardeshi, L., Somayeh, H., and Hamzehgardeshi, Z. (2014). Patient involvement in heath care decision making: A review. *Iranian Red Crescent Medical Journal, 16*(1), e12454. Retrieved from http://www.ncbi.nlm.nih.gov/pmc/articles/PMC3964421/pdf/ircmj-16-12454.pdf. Accessed 22 Feb 2016.

Weiner, B. J., Lewis, M. A., & Linnan, L. A. (2008). Using organisation theory to understand the determinants of effective implementation of worksite health promotion programmes. *Health Education Research, 24*(2), 292–305.

WHO. (2006). *Opportunities for Africa's Newborns: Practical data, policy and programmatic support for newborn care in Africa.* Geneva: World Health Organisation. Retrieved from http://www.who.int/pmnch/media/publications/oanfullreport.pdf. Accessed 31 Mar 2012.

WHO. (2016). *WHO recommendation on antenatal care for a positive pregnancy experience.* Luxembourg: World Health Organisation. Retrieved from: http://apps.who.int/iris/bitstream/10665/250796/1/9789241549912-eng.pdf?ua=1. Accessed 01 Jan 2017.

Wiig, S., Storm, M., Aase, K., Gjetsen, M. T., Solheim, M., Harthug, S., Roberts, G., Fulop, N., & QUASER Team. (2014). Investigating the use of patient involvement and patient experience in quality improvement in Norway: Rhetoric or reality? *BMC Health Services Research, 13,* 206. Retrieved from http://bmchealthservres.biomedcentral.com/articles/10.1186/1472-6963-13-206. Accessed 22 Feb 2016.

Yin, R. K. (2003). *Case study research: Design and methods* (3rd ed.). Thousand Oaks, CA: Sage.

17

How Does an Accreditation Programme in Residential Aged Care Inform the Way Residents Manage Their Healthcare and Lifestyle?

Anne Hogden, David Greenfield, Mark Brandon,
Deborah Debono, Virginia Mumford,
Johanna Westbrook and Jeffrey Braithwaite

Introduction

External regulation programmes, such as health service accreditation programmes, are designed to encourage organisations to meet standards, attain and sustain improvements, and spread lessons from which

A. Hogden (✉)
Australian Institute of Health Innovation, Macquarie University,
NSW, Australia
e-mail: anne.hogden@mq.edu.au

D. Greenfield
Australian Institute of Health Service Management,
University of Tasmania, NSW, Australia
e-mail: david.greenfield@utas.edu.au

M. Brandon
School of Business, The University of Notre Dame Australia,
Sydney, NSW, Australia

© The Author(s) 2018
A.M. McDermott et al. (eds.), *Managing Improvement in Healthcare*,
Organizational Behaviour in Health Care,
https://doi.org/10.1007/978-3-319-62235-4_17

others can learn. In many countries, the aged care sector is increasingly subject to government supervision, with regulatory requirements and performance linked to funding. Quality of care is reviewed, promoted and maintained by external assessment of residential aged care facilities, also known as nursing homes (Briggs 2006; Grenade and Boldy 2002; Hampel and Hastings 1993).

Aged care accreditation and regulation programmes internationally are similar in structure and approach. Assessors who are trained in evaluating facilities against external standards use audit methodology to review facility documentation, observe resident and staff interactions, and interview staff, residents, and their families and friends. A report detailing the outcomes of the assessment is compiled after the visit.

In Australia, these reports are publicly available. Participation in accreditation assessment is not mandatory; however, residential aged care facilities are required to be accredited to receive government subsidies. The performance of facilities is assessed against four standards: Standard 1: Management systems, staffing and organisational development; Standard 2: Resident health and personal care; Standard 3: Resident lifestyle; and Standard 4: Physical environment and safe systems. The four standards contain forty-four expected outcomes. Resident opinion is sought in relation to the outcomes. However, resident satisfaction is not directly measured, and is not reported beyond the site assessment.

D. Debono
Centre for Health Services Management, Faculty of Health,
University of Technology Sydney, Sydney, Australia

V. Mumford · J. Westbrook · J. Braithwaite
Australian Institute of Health Innovation, Macquarie University,
NSW, Australia

The quality of life of residents of aged care facilities is an important issue to residents, families, care providers, healthcare regulators and the broader community (Braithwaite 2001; Chao and Roth 2005; Chou et al. 2003; Commonwealth of Australia 2007; Hasson and Arnetz 2011; Hinchcliff et al. 2013; Street and Burge 2012). For the purposes of this chapter, the construct of 'quality of life' incorporates the concepts of resident well-being (Street et al. 2007), and resident satisfaction; that is, the meeting of residents' care and lifestyle expectations (Boldy et al. 2004; Commonwealth of Australia 2007). Consumer satisfaction with residential aged care is influenced by a range of social and environmental factors, including: social inclusion (Knight and Mellor 2007); strong social relationships within the facility, including with nursing staff (Chao and Roth 2005; Street and Burge 2012; Street et al. 2007); a sense that the facility feels 'like home' (Knight and Mellor 2007; Nakrem et al. 2013; Street et al. 2007); and a physical environment that promotes social inclusion whilst allowing respect for privacy (Chou et al. 2002a; Chou et al. 2003; Street et al. 2007). Autonomy and independence (Hillcoat-Nalletamby 2014) alter with the transition into residential aged care (Street et al. 2007). A sense of control over the transition from home to residential care (Street and Burge 2012) and the ability to exercise personal choice over aspects of daily life adds to residents' well-being (Street et al. 2007).

Residents' perceptions of a high quality of life are also influenced by organisational and staffing factors (Chou et al. 2002b; King et al. 2012; Street et al. 2007). Higher levels of nursing staff satisfaction are related to improved resident satisfaction (Chou et al. 2003; Mittal et al. 2007), suggesting that efforts made by aged care management in supporting and developing staff teams may pay dividends in improving residents' quality of life (Mittal et al. 2007). Furthermore, staff behaviour, quality of care and professional skills affect resident and family satisfaction (Chao and Roth 2005; Chou et al. 2002a; Hasson and Arnetz 2011).

Thus, there is convergence of interests of government, aged care organisations and residents and their families in residential facilities being of high quality and safety. However, there is little understanding of residents' perspectives and roles in relation to accreditation of aged care facilities. Health service accreditation research includes acute

(Braithwaite et al. 2010; Shaw et al. 2013), primary (Auras and Geraedts 2010) and aged care (Grenade and Boldy 2002) sectors, but has traditionally focused on the acute care sector (Hinchcliff et al. 2012). Our understanding of the impact of accreditation programmes in the aged care sector is limited by the lack of published research (Greenfield and Braithwaite 2008; Greenfield et al. 2013; Hinchcliff et al. 2012). The views of residents are needed to provide insight into factors influencing resident quality of life, and the relationship these bear to accreditation programmes. It is unclear if links between accreditation standards and quality of life in residential aged care are experienced or recognised by residents. Hence, the overarching question we sought to address is: how does an accreditation programme inform residents and families to manage their healthcare and lifestyle in residential aged care?

Methods

A purposive sample (Liamputtong 2009) of eleven accredited residential aged care facilities from five provider groups took part in the study. The sample was diverse in geographic location, facility size and resident care services, with sites located across four Australian states (New South Wales, South Australia, Western Australia and Queensland) in metropolitan, regional and rural areas. Participants were 71 residents living in accredited aged care facilities, most of whom were female (77%). All participants were able to contribute to discussions of their experiences; for example, one resident with communication difficulties took part with the assistance of her spouse.

We conducted focus groups (Liamputtong 2009) using a semi-structured question guide, from October 2013 to April 2014. Our questions were informed by the aged care Accreditation Standards and expected outcomes for residents' quality of life; accreditation and aged care quality of life literature; previous research on accreditation (Braithwaite et al. 2010; Braithwaite et al. 2006; Greenfield et al. 2011; Hinchcliff et al. 2013); discussions with the aged care accreditation agency; and the ongoing Australian health reforms (ACSQHC 2012). Information sought to address the research question included: residents' experiences of moving into the aged care facility; residents' perceptions of quality of care and services in residential aged care; and residents' understanding of how, or if, accreditation contributes to their care, services and quality of life.

Recordings of the focus groups were transcribed, and analysed using a bottom-up, inductive process (Thomas 2006). Excerpts addressing the research question were selected and coded for meaning (Liamputtong 2009). Examples of codes were: 'access to family'; 'residents develop relationships with long-term staff'; and 'limited availability of high-level care determined choice of facility'. Broad themes emerged through an iterative process as the codes were grouped into categories (Braun and Clarke 2006), and de-identified quotes representing resident perspectives were selected.

Results

Three themes captured residents' perceptions of accreditation standards, and how these informed their choices for their healthcare and lifestyle. These themes were: choosing a new home; adjusting to residential aged care; and supporting residents' quality of care.

Theme 1: Choosing a New Home

Accreditation information was not reported to contribute to residents' understanding of the quality of residential aged care. Residents did not make use of accreditation information to inform their choices. Rather, they expressed the view that accreditation was a pre-condition that ensured facilities offered acceptable standards of care and service. Some residents stated that they were aware of an accreditation process undertaken by facilities, and understood that the results of assessments were publicly available. Only one resident reported accessing accreditation reports when searching for an aged care facility. Whilst the reputation of the facility was considered important, the accreditation status was of limited interest, and residents saw no need to choose between facilities based on accreditation information. One resident observed that accreditation did not ensure a facility was a good place to live.

Many residents considered information provided by the Aged Care Assessment Team, an agency assessing older persons' healthcare needs, sufficient to choose a facility, as '...aged care only recommend accredited places' (D8). As residents did not access accreditation information, assessment results were not a factor influencing residents' choice of facility. A number of residents stated that they had no choice in the facility, due to their location, care needs or length of waiting lists. Accreditation information lacked relevance to residents who were unable to choose their new home.

> I think the pressure came from the hospital. They didn't ask, they just said, 'We'll send you to [facility name]', and that's all there was to it. (E10)

Some residents deferred to family members to check the accreditation assessment of the facility, whilst two residents identified that the task of checking accreditation information was too onerous for them when acutely unwell prior to admission. Family members gave support by reading through available information.

As well as reading accreditation information, family members exerted considerable influence over the selection of a facility. Most residents stated they were able to choose the home they now lived in, and identified clear reasons for their decision. Two married couples who were residents in the same facility reported choosing their facility because they could be accommodated together as a couple. Even so, the choice was frequently left to family members, who investigated the available options on residents' behalf. When making a decision, the role and views of family were paramount.

> My daughter went everywhere when she was looking, to quite a few places, and she eventually came here, and she said, 'Mum, you're going to a nice place.' She said, 'I know you don't like to leave your home, but it's time for you to go somewhere where you can be looked after, because I can't be here at all times.' (B4)

Additionally, several residents reported prior experience with the facility, either through visiting friends and family, or from volunteer work, as the reason for their choice. Familiarity with nursing staff, residents and the environment had facilitated their decision, and was considered more important than the accreditation result the facility had attained.

Theme 2: Adjusting to Residential Aged Care

The aged care accreditation standards are used to evaluate aspects of residents' lifestyle (Standard 3: Resident lifestyle). Three outcomes of this standard were considered particularly pertinent to resident adjustment. They were: emotional support (Expected outcome 3.4); privacy and dignity (Expected outcome 3.6); and choice and decision-making (Expected outcome 3.9). Residents did not make direct links between these outcomes and accreditation standards. Nonetheless, they described ways in which the expected outcomes had been enacted within the facility, and supported residents' adjustment to life in residential aged care. These actions helped residents attain an improved quality of life within their new home.

Emotional support (Expected outcome 3.4) was considered by residents to be pivotal to their adjustment. Participants acknowledged the role of staff in their transition into residential care. Many residents articulated difficulty accepting their changed situation. They related experiences of distress at the loss of their possessions and limitations to their independence and lifestyle.

> It was pretty hard to deal with when you had so many changes that you're almost broken. (D9)

Even so, most respondents identified a process of acceptance of their new life. Two factors contributed to resident perceptions of adjustment: the sense that the facility felt like home; and the development of good relationships with staff and other residents. Residents who articulated satisfaction with their new life described a sense of belonging to the facility, defined by one resident as 'a home [away] from home' (C6).

Replication of aspects of their previous life was a source of satisfaction for some participants. For resident spouse couples, the sense of home was reinforced by continuing to share their accommodation and lifestyle, despite their changing healthcare needs.

Each facility's atmosphere added to or detracted from a perception of 'hominess'. Residents described a pleasant or friendly atmosphere as contributing to their ability to settle in. The welcome received on arrival was significant to residents' adjustment. Respondents who felt welcomed by nursing staff and residents into the facility reported making efforts to welcome other new residents in their turn.

> When anyone new arrives … I always make them welcome, tell them who I am, and guide them a bit the first week or so they're here, make them feel at home. (B4)

Resident adjustment was enhanced by good relationships within the facility. Residents viewed the development of friendships with nursing staff as a key in the process of coming to terms with their new situation. Moreover, residents held the view that the size and layout of the aged care facility influenced the establishment of relationships. Smaller facilities were believed to be more conducive to the development of new friendships.

> It's got a character about it. It's the size that enables people to really feel as if they know each other. (C7)

The design of the facility's built environment contributed to respondents' perceptions of a homelike atmosphere. It could support or restrict socialisation. Residents living in multistorey facilities reported difficulty making friends on floors other than their own. One group commented that the research interview was the first opportunity to meet some of the other residents, despite having lived in the same building for several years.

Privacy and dignity, expected outcomes of Standard 3, were important issues for many residents. Being able to maintain privacy reinforced residents' sense of being at home. Most stated that their privacy and dignity was respected by staff. Even so, some considered that other residents did not always respect their privacy. Residents valued being able to spend time alone. Being able to withdraw from other residents and lock their door gave them a feeling of control over their own space: 'That's my little private place. If I want to mix I go to the community room, because I've got to have somewhere I can run away to' (B3). Additionally, sharing private time and space with family was important: 'We're able to, whether the door's open or not, have a bit of a hug' (C7).

Choice and decision-making, contained in Expected outcome 3.9, were perceived as important aspects of resident healthcare and lifestyle. Respondents reported that choice in daily care and lifestyle enhanced their quality of life. Involvement in developing their own care plans, and choices of activities and food added to residents' satisfaction: 'Everything you are asking today is gone over in the [family conference]' (F11). Whilst residents acknowledged that not everyone's wishes could be accommodated in communal life, they appreciated that nursing staff care focussed on their needs and preferences.

> ... like all the other ladies said, you couldn't wish for a nicer place to be in. Everybody's very friendly and they go to a lot of trouble for us, to make us comfortable and make our entertainment enjoyable. (B4)

Conversely, some residents reported lacking involvement in care or life-style discussions and decisions. A small number expressed dissatisfaction with limited choices available to them, and enforced changes to their behaviour: 'The rules are difficult. I am a smoker and they make me wear an apron' (A2). At times, the duty of care requirements of the organisation took precedence over the preferences of individuals. Residents were expected to accept the rules, rather than have their preferences accommodated within the bounds of safety regulation. Adjustment to residential life was hampered when care management was seen to put the requirements of the institution ahead of residents' individual needs. Attaining a better quality of life was more challenging when institutional priorities were seen to outweigh those of residents.

Theme 3: Supporting Residents' Quality of Care

Residents' views of quality of care reflected their individual experiences, and broadly aligned with the principles of the accreditation standards. Residents did not explicitly relate the standards to their perceptions of quality of care. However, they made associations between the quality of care and services they received and their sense of well-being. For example, many residents identified tangible improvements in their health and lifestyle since they had come to live in residential aged care.

> Now I'm fine, in fact I've never been better. My friends say that I look so much better, and I'm not having panic attacks, because I know I'm not alone. At home, I'd want to do something and then I'd think 'Oh, I can't go, I might pass out in the shop'… and I was a mess, but I've really, really calmed down here, and I've started to go to the shops on my own and … you get good meals. (B3)

Additionally, quality of care was perceived to be directly attributable to individual and collective staff capacity, resources and skills. Residents valued staff skills and training for residents with dementia:

> They can get a bit aggressive sometimes. I'm amazed that some of the staff are quite young and they handle it very, very well. (B3)

Relationships with staff influenced residents' perceptions of their quality of care. Residents' satisfaction was linked to a sense that they were well cared for, and they expressed a preference for permanent direct care staff: 'Permanent staff are good' (A2). Residents considered that quality of care was compromised by changes to the facility workforce. This included frequent use of casual staff, rotations and changes to staff routines.

> They have changed the staff around, I don't know what the reason is, but I think in one way it's not a good idea … they knew my likes, the ones that were here. (D9)

Stability of staffing was vital in the development of positive relationships between residents and staff. Residents' trust derived from nursing staff understanding their individual needs, and their ability and capacity to provide consistent routine care. Quality of life was sustained through improvements in health and well-being, underpinned by positive relationships with care staff.

Discussion

We examined how accreditation informs the way Australian aged care facility residents manage their healthcare and lifestyle in residential aged care. The relationship between residents' quality of life and accreditation has not been previously explored. Our study reveals that a relationship exists, but it is not explicit to residents in a way that is useful to them. Resident perceptions of a good quality of life derived from their sense of feeling at home in the aged care facility (Nakrem et al. 2013; Torrington 2007), the development of relationships (Street and Burge 2012), and the quality of care and services they received. Location, a desire to be near family and friends, and the influence of family members determined facility selection. Residents' expectation that accreditation ensured standards of quality and safety in residential aged care facilities meant that few residents made use of accreditation assessment information. Accreditation information had a limited, if any, explicit or direct influence on choice of a facility. These findings align with previous studies of resident satisfaction and quality of life (Knight and Mellor 2007) and facility selection (Cheek and Ballantyne 2001; Ryan and McKenna 2013).

Expected outcomes of Accreditation Standard 2: Health and personal care, and Standard 3: Resident lifestyle underpinned aspects of care that were important to residents. Outcomes from these standards could be linked to attaining and sustaining improved quality of life for residents. The importance of these aspects to residents was demonstrated in earlier aged care studies (Hillcoat-Nalletamby 2014; Oosterveld-Vlug et al. 2013). Despite consumer desire for the best possible quality of life, there was limited direct interest in how accreditation standards could promote a good quality of life in residential aged care. Residents' low interest in the role of accreditation is reflective of the information asymmetry existing between health providers and consumers (Grabowski and

Town 2011; Retchin 2007). A convergence of terminology between what residents seek and what accreditation assesses could help bridge this gap. Standards that are expressed in terms of partnerships with residents and families could act to promote stronger care partnerships between residents, families and staff and may better enable residents and families to negotiate areas of conflict between quality of life and health and organisational rules. Providing information on aged care facilities via 'report cards', quality ratings (Netten et al. 2012) or making accreditation reports available on the web as occurs in Australia, appears to have little impact on consumer demand (Grabowski and Town 2011). However, awareness of, and familiarity with, accreditation standards may empower residents and families to advocate for quality of life, and promote a resident-centred focus of care (Briggs 2006).

The accreditation programme has a mediating role between promoting an organisational focus on quality and safe residential care, and aspects of improvement and quality of life that are important to residents. Implicit links between accreditation and resident quality of life lead to an opportunity for increased clarity. Resident satisfaction could play an important role in refining accreditation programmes, to ensure they link directly with resident and public expectations. In Australia, resident satisfaction is assessed by surveys within aged care facilities, and directly during accreditation site visits. Formalising this within accreditation programmes and standards, and publicising this to residents and their families, would promote transparency and engagement. In doing so, accreditation programmes would: explicitly link resident satisfaction or resident priorities to drive changes in quality and safety of services; and have a greater and more explicit role to play in measuring and stimulating resident quality of life.

A further consideration arises. How might resident and family views influence the ongoing development of the accreditation programme for the aged care sector? Stronger alignment between resident priorities for quality of life and accreditation standards would be of benefit to consumers, providers and regulators. Potentially, this would allow regulators to ascertain how well standards reflect community expectations. This creates an opportunity for resident experiences to inform development of aged care accreditation programmes and standards, to disperse improvements relevant to residents and families.

Conclusion

Respect for residents' choices, privacy and dignity whilst maintaining high quality of care and safety is an ongoing issue for aged care providers and staff. Moreover, translating into standard aspects of care and service that are a priority to residents, such as the sense of being at home, or of being cared for, is a challenge for regulators and policymakers. There are opportunities for greater engagement and participation of residents in external regulation programmes to improve their efficacy.

References

ACSQHC. (2012). *Australian commission on safety and quality in health care: Australian safety and quality goals for health care: Development and consultation report.* Sydney: Australian Commission on Safety and Quality in Health Care.

Auras, S., & Geraedts, M. (2010). Patient experience data in practice accreditation—An international comparison. *International Journal for Quality in Health Care, 22,* 132–139. doi:10.1093/intqhc/mzq006mzq006.

Boldy, D. P., Chou, S. C., & Lee, A. H. (2004). Assessing resident satisfaction and its relationship to staff satisfaction in residential aged care. *Australasian Journal of Ageing, 23,* 195–197. doi:10.1111/j.1741-6612.2004.00049.x.

Braithwaite, J. (2001). Regulating nursing homes: The challenge of regulating care for older people in Australia. *BMJ, 323,* 443–446.

Braithwaite, J., et al. (2006). A prospective, multi-method, multi-disciplinary, multi-level, collaborative, social-organisational design for researching health sector accreditation. *BMC Health Services Research, 6,* 113. doi:10.1186/1472-6963-6-113.

Braithwaite, J., et al. (2010). Health service accreditation as a predictor of clinical and organisational performance: A blinded, random, stratified study. *Quality and Safety in Health Care, 19,* 14–21. doi:10.1136/qshc.2009.033928.

Braun, V., & Clarke, C. (2006). Using thematic analysis in psychology. *Qualitative Research Psychology, 3,* 77–101. doi:10.1191/1478088706qp063oa.

Briggs, D. (2006). Accreditation and the quality journey in aged care. *Asian Journal of Gerontology and Geriatrics, 1,* 163–169.

Chao, S. Y., & Roth, P. (2005). Dimensions of quality in long-term care facilities in Taiwan. *Journal of Advanced Nursing, 52,* 609–618. doi:10.1111/j.1365-2648.2005.03632.x.

Cheek, J., & Ballantyne, A. (2001). Moving them on and in: The process of searching for and selecting an aged care facility. *Qualitative Health Research, 11,* 221–237.

Chou, S. C., Boldy, D. P., & Lee, A. H. (2002a). Resident satisfaction and its components in residential aged care. *The Gerontologist, 42,* 188–198. doi:10.1093/geront/42.2.188.

Chou, S. C., Boldy, D. P., & Lee, A. H. (2002b). Staff satisfaction and its components in residential aged care. *International Journal for Quality in Health Care, 14,* 207–217. doi:10.1093/oxfordjournals.intqhc.a002612.

Chou, S. C., Boldy, D. P., & Lee, A. H. (2003). Factors influencing residents' satisfaction in residential aged care. *The Gerontologist, 43,* 459–472. doi:10.1093/geront/43.4.459.

Commonwealth of Australia. (2007). *Evaluation of the impact of accreditation on the delivery of quality of care and quality of life to residents in Australian Government subsidised residential aged care homes.* Canberra: Australia.

Grabowski, D. C., & Town, R. J. (2011). Does information matter? Competition, quality, and the impact of nursing home report cards. *Health Services Research, 46,* 1698–1719. doi:10.1111/j.1475-6773.2011.01298.x.

Greenfield, D., & Braithwaite, J. (2008). Health sector accreditation research: A systematic review. *International Journal for Quality in Health Care, 20,* 172–183. doi:10.1093/intqhc/mzn005mzn005.

Greenfield, D., Pawsey, M., & Braithwaite, J. (2011). What motivates professionals to engage in the accreditation of healthcare organizations? *International Journal of Quality in Health Care, 23,* 8–14. doi:10.1093/intqhc/mzq069.

Greenfield, D., Pawsey, M., & Braithwaite, J. (2013). Accreditation: A global regulatory mechanism to promote quality and safety. In W. Sollecito & J. Johnson (Eds.), *Continuous quality improvement in health care* (4th ed., pp. 513–531). New York: Jones and Barlett Learning.

Grenade, L., & Boldy, D. (2002). The accreditation experience: Views of residential aged care providers. *Geriaction, 20,* 5–9.

Hampel, M. J., & Hastings, M. M. (1993). Assessing quality in nursing home dementia special care units: A pilot test of the joint commission protocol. *Journal of Mental Health Administration, 20,* 236–246. doi:10.1007/bf02518692.

Hasson, H., & Arnetz, J. E. (2011). Care recipients' and family members' perceptions of quality of older people care: A comparison of home-based care and nursing homes. *Journal of Clinical Nursing, 20,* 1423–1435. doi:10.1111/j.1365-2702.2010.03469.x.

Hillcoat-Nalletamby, S. (2014). The meaning of "independence" for older people in different residential settings. *The Journals of Gerontology, Series B: Psychological Sciences and Social Sciences, 69,* 419–430. doi:10.1093/geronb/gbu008.

Hinchcliff, R., Greenfield, D., Moldovan, M., Westbrook, J. I., Pawsey, M., Mumford, V., & Braithwaite, J. (2012). Narrative synthesis of health service accreditation literature. *BMJ Quality and Safety, 21,* 979–991. doi:10.1136/bmjqs-2012-000852bmjqs-2012-000852.

Hinchcliff, R., Greenfield, D., Westbrook, J. I., Pawsey, M., Mumford, V., & Braithwaite, J. (2013). Stakeholder perspectives on implementing accreditation programs: A qualitative study of enabling factors. *BMC Health Services Research, 13,* 437. doi:10.1186/1472-6963-13-437.

King, D., et al. (2012). *The aged care workforce.* Canberra: Australia.

Knight, T., & Mellor, D. (2007). Social inclusion of older adults in care: Is it just a question of providing activities? *International Journal of Qualitative Studies on Health and Well-being, 2,* 76–85. doi:10.1080/17482620701320802.

Liamputtong, P. (2009). *Qualitative research methods* (3rd ed.). Melbourne: Oxford University Press.

Mittal, V., et al. (2007). Perception gap in quality-of-life ratings: An empirical investigation of nursing home residents and caregivers. *The Gerontologist, 47,* 159–168. doi:10.1093/geront/47.2.159.

Nakrem, S., Vinsnes, A. G., Harkless, G. E., Paulsen, B., & Seim, A. (2013). Ambiguities: Residents' experience of 'nursing home as my home'. *International Journal of Older People Nursing, 8,* 216–225. doi:10.1111/j.1748-3743.2012.00320.x.

Netten, A., Trukeschitz, B., Beadle-Brown, J., Forder, J., Towers, A. M., & Welch, E. (2012). Quality of life outcomes for residents and quality ratings of care homes: Is there a relationship? *Age and Ageing, 41,* 512–517. doi:10.1093/ageing/afs050.

Oosterveld-Vlug, M. G., Pasman, H. R., van Gennip, I. E., Muller, M. T., Willems, D. L., & Onwuteaka-Philipsen, B. D. (2013). Dignity and the factors that influence it according to nursing home residents: A qualitative interview study. *Journal of Advanced Nursing, 70*(1), 97–106. doi:10.1111/jan.12171.

Retchin, S. M. (2007). Overcoming information asymmetry in consumer-directed health plans. *American Journal of Managed Care, 13*, 173–176.

Ryan, A., & McKenna, H. (2013). 'Familiarity' as a key factor influencing rural family carers' experience of the nursing home placement of an older relative: A qualitative study. *BMC Health Services Research, 13*, 252. doi:10.1186/1472-6963-13-252.

Shaw, C. D., Braithwaite, J., Moldovan, M., Nicklin, W., Grgic, I., Fortune, T., & Whittaker, S. (2013). Profiling health-care accreditation organizations: An international survey. *International Journal of Quality in Health Care, 25*, 222–231. doi:10.1093/intqhc/mzt011.

Street, D., & Burge, S. (2012). Residential context, social relationships, and subjective well-being in assisted living. *Research on Aging, 34*, 365–394. doi:10.1177/0164027511423928.

Street, D., Burge, S., Quadagno, J., & Barrett, A. (2007). The salience of social relationships for resident well-being in assisted living. *The Journals of Gerontology: Series B, Psychological Sciences and Social Sciences, 62*, S129–S134. doi:10.1093/geronb/62.2.s129.

Thomas, D. R. (2006). A general inductive approach for analyzing qualitative evaluation data. *American Journal of Evaluation, 27*, 237–246. doi:10.1177/1098214005283748.

Torrington, J. (2007). Evaluating quality of life in residential care buildings. *Building Research & Information, 35*, 514–528. doi:10.1080/09613210701318102.

Index

© The Editor(s) (if applicable) and The Author(s) 2018
A.M. McDermott et al. (eds.), *Managing Improvement in Healthcare*,
Organizational Behaviour in Health Care,
https://doi.org/10.1007/978-3-319-62235-4

CPSIA information can be obtained
at www.ICGtesting.com
Printed in the USA
LVOW13*2331161017
552621LV00019B/361/P